MW00837524

RARE EARTH MINERALS: POLICIES AND ISSUES

EARTH SCIENCES IN THE 21ST CENTURY

Additional books in this series can be found on Nova's website
under the Series tab.

Additional E-books in this series can be found on Nova's website
under the E-books tab.

MATERIALS SCIENCE
AND TECHNOLOGIES

Additional books in this series can be found on Nova's website
under the Series tab.

Additional E-books in this series can be found on Nova's website
under the E-books tab.

EARTH SCIENCES IN THE 21ST CENTURY

RARE EARTH MINERALS: POLICIES AND ISSUES

STEVEN M. FRANKS
EDITOR

Nova Science Publishers, Inc.
New York

For permission to use material from this book please contact us:
Telephone 631-231-7269; Fax 631-231-8175
Web Site: http://www.novapublishers.com

NOTICE TO THE READER

Additional color graphics may be available in the e-book version of this book.

LIBRARY OF CONGRESS CATALOGING-IN-PUBLICATION DATA

Rare earth minerals : policies and issues / editor, Steven M. Franks.
    p. cm.
  Includes index.
  ISBN 978-1-61122-310-1 (hardcover)
  1. Rare earths. 2. Rare earth metalls. 3. Rare earth alloys. 4.
Business logistics--United States. I. Franks, Steven M.
  QE390.2.R37R36 2010
  338.2'7494--dc22
                    2010041601

Published by Nova Science Publishers, Inc. ✛ New York

CONTENTS

PREFACE

U.S. mineral policy emphasizes developing domestic supplies of critical materials and the domestic private sector to produce and process those materials. However, some raw materials do not exist in economic quantities in the United States, while processing, manufacturing, and other downstream ventures in the United States may not be competitive with facilities in other regions of the world. The current goal of U.S. mineral policy is to promote an adequate, stable, and reliable supply of materials for U.S. national security, economic well-being, and industrial production. This book provides a discussion on the major issues and concerns of the global supply chain for rare earth elements, their major end uses, and legislative and other policy proposals that Congress may consider to improve the U.S. rare earth position.

Chapter 1- The concentration of production of rare earth elements (REEs) outside the United States raises the important issue of supply vulnerability. REEs are used for new energy technologies and national security applications. Is the United States vulnerable to supply disruptions of REEs? Are these elements essential to U.S. national security and economic well-being?

There are 17 rare earth elements (REEs), 15 within the chemical group called lanthanides, plus yttrium and scandium. The lanthanides consist of the following: lanthanum, cerium, praseodymium, neodymium, promethium, samarium, europium, gadolinium, terbium, dysprosium, holmium, erbium, thulium, ytterbium, and lutetium. Rare earths are moderately abundant in the earth's crust, some even more abundant than copper, lead, gold, and platinum. While more abundant than many other minerals, REE are not concentrated enough to make them easily exploitable economically. The United States was once self-reliant in domestically produced REEs, but over the past 15 years has become 100% reliant on imports, primarily from China, because of lower-cost operations.

Chapter 2- This letter formally transmits the enclosed briefing in response to the National Defense Authorization Act for Fiscal Year 2010 (Pub. L. No. 111-84), which required GAO to submit a report on rare earth materials in the defense supply chain to the Committees on Armed Services of the Senate and House of Representatives by April 1, 2010. As required, we provided a copy of this briefing to the committees on April 1, 2010, and subsequently briefed the Senate Armed Services Committee staff on April 5, 2010, and the House Armed Services Committee staff on April 6, 2010.

We are sending copies of this report to the appropriate congressional committees. We are also sending copies to the Secretaries of Defense, Commerce, Energy, and the Interior. This report is also available at no charge on the GAO Web site at http://www.gao.gov. Should you or your staff have any questions concerning this report, please contact me at (202) 512-4841 or martinb@gao.gov. Contact points for our Offices of Congressional Relations and Public Affairs may be found on the last page of this report. Key contributors to this report were John Neumann, Assistant Director; James Kim; Erin Carson; Brent Corby; Marie Ahearn; Barbara El Osta; and Morgan Delaney Ramaker.

Chapter 3- I'd like to thank Chairman Miller for calling this hearing. Last September, I saw an article on this issue that raised a number of questions in my mind about whether the Committee and the Congress were doing enough to support American business and American jobs.

Rare earths are an essential component in a wide array of emerging industries.

This is not the first time the Committee has been concerned with the competitive implications of materials such as rare earths. In 1980—30 years ago—this Committee established a national minerals and materials policy. One core element in that legislation was the call to support for "a vigorous, comprehensive and coordinated program of materials research and development."

Chapter 4- Welcome to our hearing this afternoon on something most of us have never heard of at all, or promptly forgot after our test on the Periodic Table in high school chemistry. Today we will be discussing rare earth elements, which aren't really all that rare. Rare earth elements are crucial to making the magnets and batteries needed for the energy industry of the 21st Century. With a little of one of these elements you can get a smaller, more powerful magnet, or an aircraft engine that operates at higher temperatures or a fiber-optic cable that can carry your phone call much greater distances.

Chapter 5- Mineral-based materials are ubiquitous—aluminum in jet aircraft; steel in bridges and buildings, and lead in batteries, to name but a few examples. The emergence of new technologies and engineered materials creates the prospect of rapid increases in demand for some minerals previously used in relatively small quantities in a small number of applications—such as lithium in automotive batteries, rare-earth elements in permanent magnets and compact-fluorescent light bulbs, and indium and tellurium in photovoltaic solar cells. At the same time, the supplies of some minerals seemingly are becoming increasingly fragile due to more fragmented supply chains, increased U.S. import dependence, export restrictions by some nations on primary raw materials, and increased industry concentration

Chapter 6- The Ames Laboratory (AL) is the smallest of DOE's (Department of Energy) seventeen national laboratories. It is a single-program laboratory with about 78% of DOE's funding from the Office of Basic Energy Sciences (BE S). Additional non-DOE income of about $7M is derived from contracts, grants, Cooperative Research and Development Agreements (CRADAs) and Work for Others (WFO) arrangements.

The AL is fully integrated with Iowa State University (ISU) and its buildings are right on the campus and several are directly connected with ISU buildings. All AL personnel are ISU employees, and many of the lead scientists (23) have joint appointments with various academic departments. There are about 140 scientists and engineers, 260 graduate and undergraduate students, another 240 visiting scientists, facility users, and associates, and 180 support personnel, for a total of about 440 employees (or about 300 full-time equivalent employees) and 410 associates (non-payroll). Of the scientific staff about 20% are directly involved in rare earth research and development activities, including materials science, condensed matter physics, and materials chemistry.

Chapter 7- GE is a diversified global infrastructure, finance, and media company that provides a wide array of products to meet the world's essential needs. From energy and water to transportation and healthcare, we are driving advanced technology and product solutions in key industries central to providing a cleaner, more sustainable future for our nation and the world.

At the core of every GE product are the materials that make up that product. To put GE's material usage in perspective, we use at least 70 of the first 83 elements listed in the Periodic Table of Elements. In actual dollars, we spend $40 billion annually on materials. 10% of this is for the direct purchase of metals and alloys. In the specific case of the rare earth elements, we use these elements in our Healthcare, Lighting, Energy, Motors, and Transportation products.

Chapter 8- I'm the CEO of rare earths technology company Molycorp Minerals, LLC. Molycorp owns the rare earth mine and processing facility at Mountain Pass, California, one of the richest rare earth deposits in the world, and we are the only active producer of rare earths in the Western Hemisphere. I have worked with Molycorp and its former parent companies, Unocal and Chevron, for over 25 years, and have watched closely the evolution of this industry over the past decade. It has been remarkable to watch the applications for rare earths explode. However, as rare earth-based technologies have become more and more essential, the U.S., which invented rare earth processing and manufacturing technology, has become almost completely dependent on China for access to rare earths and, more specifically, the metals, alloys and magnets that derive from them.

Chapter 9- Let me start with some acknowledgments on my limitations as a witness on rare earth minerals. First, my background and expertise is on international trade law matters, including the World Trade Organization, and manufacturing competitiveness issues. Others on the panel today are the experts on minerals in general or rare earth minerals policies.

Our firm, over the years, has looked at many aspects of the U.S.-China relationship and has prepared for the U.S.-China Economic and Security Review Commission various studies looking at the trade and manufacturing impacts of China's practices. For example, on March 24, 2009 I testified at a hearing before the Commission on "China's Industrial Policy and its Impact on U.S. Companies, Workers and the American Economy."

Chapter 10-: In 2009, rare earths were not mined in the United States; however, rare-earth concentrates previously produced at Mountain Pass, CA, were processed into lanthanum concentrate and didymium (75% neodymium, 25% praseodymium) products. Rare-earth concentrates, intermediate compounds, and individual oxides were available from stocks. The United States continued to be a major consumer, exporter, and importer of rare-earth products in 2009. The estimated value of refined rare earths imported by the United States in 2009 was $84 million, a decrease from $186 million imported in 2008. Based on final 2008 reported data, the estimated 2008 distribution of rare earths by end use, in decreasing order, was as follows: metallurgical applications and alloys, 29%; electronics, 18%; chemical catalysts, 14%; rare-earth phosphors for computer monitors, lighting, radar, televisions, and x-ray-intensifying film, 12%;

automotive catalytic converters, 9%; glass polishing and ceramics, 6%; permanent magnets, 5%; petroleum refining catalysts, 4%; and other, 3%.

Chapter 11- The rare-earth element yttrium was not mined in the United States in 2009. All yttrium metal and compounds used in the United States were imported. Principal uses were in phosphors for color televisions and computer monitors, temperature sensors, trichromatic fluorescent lights, and x-ray-intensifying screens. Yttria-stabilized zirconia was used in alumina-zirconia abrasives, bearings and seals, high-temperature refractories for continuous-casting nozzles, jet-engine coatings, oxygen sensors in automobile engines, simulant gemstones, and wear-resistant and corrosion-resistant cutting tools. In electronics, yttrium-iron garnets were components in microwave radar to control high-frequency signals. Yttrium was an important component in yttrium-aluminum-garnet laser crystals used in dental and medical surgical procedures, digital communications, distance and temperature sensing, industrial cutting and welding, nonlinear optics, photochemistry, and photoluminescence. Yttrium also was used in heating-element alloys, high-temperature superconductors, and superalloys. The approximate distribution in 2008 by end use was as follows: phosphors (all types), 87%; ceramics, 10%; metallurgy, 2%; and electronics and lasers, 1 %.

In: Rare Earth Minerals: Policies and Issues   ISBN: 978-1-61122-310-1
Editor: Steven M. Franks   © 2011 Nova Science Publishers, Inc.

Chapter 1

RARE EARTH ELEMENTS: THE GLOBAL SUPPLY CHAIN

Marc Humphries

SUMMARY

The concentration of production of rare earth elements (REEs) outside the United States raises the important issue of supply vulnerability. REEs are used for new energy technologies and national security applications. Is the United States vulnerable to supply disruptions of REEs? Are these elements essential to U.S. national security and economic well-being?

There are 17 rare earth elements (REEs), 15 within the chemical group called lanthanides, plus yttrium and scandium. The lanthanides consist of the following: lanthanum, cerium, praseodymium, neodymium, promethium, samarium, europium, gadolinium, terbium, dysprosium, holmium, erbium, thulium, ytterbium, and lutetium. Rare earths are moderately abundant in the earth's crust, some even more abundant than copper, lead, gold, and platinum. While more abundant than many other minerals, REE are not concentrated enough to make them easily exploitable economically. The United States was once self-reliant in domestically produced REEs, but over the past 15 years has become 100% reliant on imports, primarily from China, because of lower-cost operations.

There is no rare earth mine production in the United States. U.S.-based Molycorp operates a separation plant at Mountain Pass, CA, and sells the rare earth concentrates and refined products from previously mined above-ground stocks. Neodymium, praseodymium, and lanthanum oxides are produced for further processing but these materials are not turned into rare earth metal in the United States.

Some of the major end uses for rare earth elements include use in automotive catalytic converters, fluid cracking catalysts in petroleum refining, phosphors in color television and flat panel displays (cell phones, portable DVDs, and laptops), permanent magnets and rechargeable batteries for hybrid and electric vehicles, and generators for wind turbines, and numerous medical devices. There are important defense applications, such as jet fighter engines, missile guidance systems, antimissile defense, and space-based satellites and communication systems.

World demand for rare earth elements is estimated at 134,000 tons per year, with global production around 124,000 tons annually. The difference is covered by previously mined aboveground stocks. World demand is projected to rise to 180,000 tons annually by 2012, while it is unlikely that new mine output will close the gap in the short term. New mining projects could easily take 10 years to reach production. In the long run, however, the USGS expects that global reserves and undiscovered resources are large enough to meet demand.

Legislative proposals H.R. 4866 (Coffman) and S. 3521(Murkowski) have been introduced to support domestic production of REEs, because of congressional concerns over access to rare earth raw materials and downstream products used in many national security applications and clean energy technologies.

INTRODUCTION

The concentration of production of rare earth elements (REEs) raises the important issue of supply vulnerability. REEs are used for new energy technologies and national security applications. Is the U.S. vulnerable to supply disruptions? Are these elements essential to U.S. national security and economic well-being?

The examination of REEs for new energy technologies reveals a concentrated and complex global supply chain and numerous end-use applications. Placing the REE supply chain in the global context is unavoidable.

U.S. mineral policy emphasizes developing domestic supplies of critical materials and the domestic private sector to produce and process those materials.[1] But some raw materials do not exist in economic quantities in the United States, while processing, manufacturing, and other downstream ventures in the United States may not be competitive with facilities in other regions of the world. However, there may be public policies enacted or executive branch measures taken to offset the U.S. disadvantage of its potentially higher cost operations. The current goal of U.S. mineral policy is to promote an adequate, stable, and reliable supply of materials for U.S. national security, economic well-being, and industrial production.

Aside from a small amount of recycling, the United States is 100% reliant on imports of REEs and highly dependent on many other minerals that support its economy. For example, the United States is more than 90% import-reliant for manganese (100%), bauxite (100%), platinum (94%), and uranium (90%). While import reliance may be a cause for concern, high import reliance is not necessarily the best measure, or even a good measure, of supply risk. The supply risk for bauxite, for example, may not be the same as that for REEs. However, in the case of REEs, the dominance of China as a single or dominant supplier of the raw material, downstream oxides, associated metals and alloys, is a cause for concern because of China's growing internal demand for its REEs.

This report provides a discussion on the major issues and concerns of the global supply chain for rare earth elements, their major end uses, and legislative and other policy proposals that Congress may consider to improve the U.S. rare earth position.

WHAT ARE RARE EARTH ELEMENTS?

There are 17 rare earth elements (REEs), 15 within the chemical group called lanthanides, plus yttrium and scandium. The lanthanides consist of the following: lanthanum, cerium, praseodymium, neodymium, promethium, samarium, europium, gadolinium, terbium, dysprosium, holmium, erbium, thulium, ytterbium, and lutetium. Rare earths are moderately abundant in the earth's crust, some even more abundant than copper, lead, gold, and platinum.

While some are more abundant than many other minerals, most REEs are not concentrated enough to make them easily exploitable economically.[2] The United States was once self-reliant in domestically produced REEs, but over the past 15 years has become 100% reliant on imports, primarily from China, because of lower-cost operations.[3]

MAJOR END USES AND APPLICATIONS

Currently, the dominant end use for rare earth elements in the U.S. are for auto catalysts and petroleum refining catalysts. Some other major end uses for rare earth elements include use in phosphors in color television and flat panel displays (cell phones, portable DVDs, and laptops), permanent magnets and rechargeable batteries for hybrid and electric vehicles, and numerous medical devices. There are important defense applications such as jet fighter engines, missile guidance systems, antimissile defense, and space-based satellites and communication systems. Permanent magnets containing neodymium, gadolinium, dysprosium, and terbium are used in numerous electrical and electronic components and generators for wind turbines. See Table 1 below for selected end uses of rare earth elements.

Table 1. Rare Earth Elements (Lanthanides): Selected End Uses

Light Rare Earths (more abundant)	Major End Use	Heavy Rare Earth (less abundant)	Major End Use
Lanthanum	hybrid engines, metal alloys	Terbium	phosphors, permanent magnets
Cerium	auto catalyst, petroleum refining, metal alloys	Dysprosium	permanent magnets, hybrid engines
Praseodymium	magnets	Erbium	phosphors
Neodymium	auto catalyst, petroleum refining, hard drives in laptops, headphones, hybrid engines	Yttrium	red color, fluorescent lamps, ceramics, metal alloy agent
Samarium	magnets	Holmium	glass coloring, lasers
Europium	red color for television and computer screens	Thulium	medical x-ray units
Gadolinium	magnets	Lutetium	catalysts in petroleum refining
		Ytterbium	lasers, steel alloys

Source: DOI, U.S. Geological Survey, Circular 930-N.

Demand for Rare Earth Elements

World demand for rare earth elements is estimated at 134,000 tons per year, with global production around 124,000 tons annually. The difference is covered by above-ground stocks or inventories. World demand is projected to rise to 180,000 tons annually by 2012, while it is unlikely that new mine output will close the gap in the short term.[4] By 2014, global demand for rare earth elements may exceed 200,000 tons per year. China's output may reach 160,000 tons per year (up from 130,000 tons in 2008) in 2014. An additional capacity shortfall of 40,000 tons per year may occur. This potential shortfall has raised concerns in the U.S. Congress. New mining projects could easily take 10 years for development. In the long run, however, the USGS expects that reserves and undiscovered resources are large enough to meet demand.

While world demand continues to climb, U.S. demand for rare earths is also projected to rise, according to the USGS Commodity Specialist Jim Hedrick.[5] For example, permanent magnet demand is expected to grow by 10%-16% per year through 2012. Demand for rare earths in auto catalysts and petroleum cracking catalysts is expected to increase between 6% and 8% each year over the same period. Demand increases are also expected for rare earths in flat panel displays, hybrid vehicle engines, and defense and medical applications.

The Application of Rare Earth Metals in National Defense[6]

Current government policies pertaining to the acquisition of certain minerals for defense purposes are addressed, in part, in several different legislative initiatives, including the Defense Production Act (P.L. 8 1-774), National Defense Stockpile [Title 50 United States Code (U.S.C.) 98-h-2(a)],[7] Buy American Act (41 U.S.C. 10-10d), Berry Amendment (10 U.S.C. 2533a), and the Specialty
Metal provision (10 U.S.C. 2533b). However, these policies do not present a unified opinion on whether every mineral is considered "critical," "strategic," or necessary for national security purposes, and there is a certain lack of cohesion to the application of these policies. As an example, rare earth elements (and rare earth metals) fall outside of the scope of the Berry Amendment and the Specialty Metal provision.[8]

The primary defense application of rare earth materials is their use in four types of permanent magnet materials commercially available: Alnico, Ferrites,

Samarium Cobalt, and Neodymium Iron Boron. With the exception of Neodymium Iron Boron, all of the materials are domestically produced. The United States has no production capabilities for Neodymium Iron Boron. Neo magnets, the product derived from Neodymium Iron Boron, and Samarium Cobalt, are considered important to many defense products. They are considered one of the world's strongest permanent magnets and an essential element to many military weapons systems, as described in the following examples.

- Jet fighter engines and other aircraft components, including samarium-cobalt magnets used in generators that produce electricity for aircraft electrical systems;
- Missile guidance systems, including precision guidance munitions, lasers, and smart bombs;[9]
- Electronic countermeasures systems;
- Underwater mine detection systems;
- Antimissile defense systems;
- Range finders, including lasers; and
- Satellite power and communication systems, including traveling wave tubes (TWT) rare earth speakers, defense system control panels, radar systems, electronic counter measures, and optical equipment.[10]

Many scientific organizations have concluded that certain rare earth metals are critical to U.S. national security and becoming increasingly more important in defense applications.[11] Some industry analysts are concerned with an increasing dependence on foreign sources for rare earth metals; a dwindling source of domestic supply for certain rare earth metals; and the emergence of a manufacturing supply chain that has largely migrated outside of the United States. In July 2010, the China Ministry of Commerce announced that China would cut its export quota for rare earth minerals by 72%, raising concerns because of estimates that China controls approximately 97% of the global production of rare earth minerals.[12] It is also estimated that by 2012 China's domestic consumption will outpace China's domestic production of rare earth minerals.

Some experts are concerned that DOD is not doing enough to mitigate the possible risk posed by a scarcity of domestic suppliers. As an example, the United States Magnet Materials Association (USMMA), a coalition of companies representing aerospace, medical, and electronic materials, has recently expanded its focus to include rare earth metals and the rare earth magnet supply chain. In February 2010, USMMA unveiled a six-point plan to address what they describe as the "impending rare earth crisis" which they assert poses a significant threat to the economy and national security of the United States.[13] However, it appears that DOD's position assumes that there are a sufficient number of supplier countries worldwide to mitigate the potential for shortages.

RARE EARTH RESOURCES AND PRODUCTION POTENTIAL

Rare earth elements often occur with other elements, such as copper, gold, uranium, phosphates, and iron, and have often been produced as a byproduct. The lighter elements such as lathananum, cerium, praseodymium, and neodymium are more abundant and concentrated and usually make up about 80%-99% of a total deposit. The heavier elements—gadolinium through lutetium and yttrium—are scarcer but very "desirable," according to USGS commodity analysts.[14]

Most rare earth elements throughout the world are located in deposits of the minerals bastnaesite[15] and monazite.[16] Bastnaesite deposits in the United States and China account for the largest concentrations of REEs, while monazite deposits in Australia, South Africa, China, Brazil, Malaysia, and India account for the second largest concentrations of REEs. Bastnaesite occurs as a primary mineral, while monazite is found in primary deposits of other ores and typically recovered as a byproduct. Over 90% of the world's economically recoverable rare earth elements are found in primary mineral deposits (i.e., in bastnaesite ores).[17]

Concerns over radioactive hazards associated with monazites (because it contains thorium) has nearly eliminated it as a REE source in the United States. Bastnaesite, a low-thorium mineral (dominated by lanthanum, cerium, and neodymium) is shipped from stocks in Mountain Pass, CA. The more desirable heavy rare earth elements account for only 0.4% of the total stock. Monazites have been produced as a minor byproduct of uranium and niobium processing. Rare earth element reserves and resources are found in Colorado, Idaho, Montana, Missouri, Utah, and Wyoming. Heavy rare earth elements (HREEs) dominate in the Quebec-Labrador (Strange Lake) and Northwest Territories (Thor Lake) areas of Canada. There are high-grade deposits in Banyan Obo, Inner Mongolia, China (where much of the world's REE production is taking place) and lower-grade deposits in South China provinces providing a major source of the heavy rare earth elements.[18] Areas considered to be attractive for

REE development include Strange Lake and Thor Lake in Canada; Karonga, Burundi; and Wigu Hill in Southern Tanzania.

Careful consideration should be given to the feasibility of mining and processing of REEs as a byproduct of phosphorus deposits and from titanium and niobium mines in Brazil.[19] Canadian, Chinese, and U.S. firms have recently assessed various REE deposits associated with development of primary minerals such as gold, iron ore, and mineral sand projects in the United States. Table 2 below illustrates China's near-monopoly position in world rare earth production. However, REE reserves and the reserve base are more dispersed throughout the world. China holds 36% of the world's reserves (36 million metric tons out of 99 million metric tons) and the United States holds about 13%. South Africa and Canada (included in the "Other" category) have significant REE potential, according to the USGS. REE reserves are also found in Australia, Brazil, India, Russia, South Africa, Malaysia, and Malawi.

Table 2. Rare Earth Elements: World Production and Reserves—2009

Country	Mine Production (metric tons)	% of total	Reserves (million metric tons)	% of total	Reserve Base[a] (million metric tons)	% of total
United States	none		13.0	13	14.0	9.3
China	120,000	97	36.0	36	89.0	59.3
Russia			19.0	19	21.0	14
(and other former Soviet Union countries)						
Australia			5.4	5	5.8	3.9
India	2,700	2	3.1	3	1.3	1
Brazil	650		small			
Malaysia	380		small			
Other	270		22.0	22	23	12.5
Total	124,000		99.0		154	

Source: U.S. Department of the Interior, Mineral Commodity Summaries, USGS, 2010.
a. Reserve Base is defined by the USGS to include reserves (both economic and marginally economic) plus some subeconomic resources (i.e., those that may have potential for becoming economic reserves).

There is no rare earth mine production in the United States. U.S.-based Molycorp operates a separation plant at Mountain Pass, CA, and sells the rare earth concentrates and refined products from previously mined above-ground stocks. Neodymium, praseodymium, and lanthanum oxides are produced for further processing, but these materials are not turned into rare earth metal in the

United States. While the United States exports much of its REE stocks to Japan, that material is not counted in the trade equation for import reliance because the material is not produced from a primary source.

Molycorp, which has an exploration program underway to further delineate its rare earth mineral deposits, has plans for full mine production in the second half of 2012 and has plans to modernize its refinery facilities. Molycorp's Mountain Pass deposit contained an estimated 30 million tons of REE reserves and once produced as much as 20,000 tons per day.[20] Mountain Pass cut-off grade (below which the deposit may be uneconomic) is, in some parts, 7.6%, while the average grade is 9.6%. U.S. Rare Earth (another U.S. based company), in the pre-feasibility stage of mine development, has long-term potential because of its large deposits in Idaho, Colorado, and Montana.[21]

Canadian deposits contain the heavy rare earth elements dysprosium, terbium, and europium, which are needed for magnets to operate at high temperatures. Great Western Minerals Group (GWMG) of Canada and Avalon Rare Metals have deposits with an estimated high content (1%- 2%) of heavy rare earth elements.[22] Avalon is developing a rare earth deposit at Thor Lake in the Northwest Territories of Canada. Drilling commenced in January 2010. Thor Lake is considered by some in the industry to contain one of the largest REE deposits in the world with the potential for production of heavy REEs.[23] GWMG owns a magnet alloy producer in the U.K. When GWMG begins production in Canada and elsewhere, they plan to have a refinery near the mine site allowing greater integration and control over the supply chain. Great Western's biggest advantage could be its potential for a vertically integrated operation.

The Lynas Corp., based in Australia, has immediate potential for light rare earths development, according to investor analyst Jack Lifton. Development of Lynas's Mt. Weld deposit in Australia is underway and there is potential to reopen the rare earth mine Steenkampskraal in South Africa. An agreement between GWMG and Rare Earth Extraction Co. Ltd. of Stellenbosch to develop the mine is in progress. The Japan Oil, Gas, and Metals National Corporation (JOGMEC) signed an agreement with Midland Exploration Inc. for development of the Ytterby project in Quebec, Canada. JOGMEC is under the authority of the Japanese Ministry of Economy, Trade, and Industry with a mandate to invest in projects worldwide to receive access to stable supplies of natural resources for Japan.

Access to a reliable supply to meet current and projected demand is an issue of concern. In 2009, China produced 97% of the world's rare earth elements (measured in rare earth oxide content) and continues to restrict exports of the material through quotas and export tariffs. China has plans to reduce mine output, eliminate illegal operations, and restrict REE exports even further. There are some immediate supply concerns with lower rare earth export quotas in China. China has cut its exports of rare earth elements from about 50,000 metric tons in 2009 to 30,000 metric tons in 2010. According to a Bloomberg news report, a July 2010 announcement by China's Ministry of Commerce would cut exports of REEs by 72%, to about 8,000 metric tons, for the second half of 2010.[24]

Further restrictions

While limited production and processing capacity for rare earths currently exists elsewhere in the world, additional capacity is expected to be developed in the United States, Australia, and Canada within two to five years, according to some experts.[25] Chinese producers are also seeking to expand their production capacity in areas around the world, particularly in Australia. There are only a few exploration companies that develop the resource, and because of long lead times needed from discovery to refined elements, supply constraints are likely in the short term.

Supply Chain Issues

The supply chain for rare earth elements generally consist of mining, separation, refining, alloying, and manufacturing (devices and component parts). A major issue for REE development in the United States is the lack of refining, alloying, and fabricating capacity that could process any future rare earth production. There are two U.S. companies (Electron Energy Corporation (EEC) in Landisville, PA, and Santoku America in Tolleson, AZ) producing samarium-cobalt (Sm-Co) permanent magnets, while there are no U.S. producers of the more desirable neodymium-iron-boron magnets needed for numerous consumer electronics, energy, and defense applications. Even the REEs needed for these magnets that operate at the highest temperatures include small amounts of dysprosium and terbium, both available only from China at the moment. EEC, in its production of its Sm-Co permanent magnet, uses small amounts of gadolinium—an REE of which there is no U.S. production. EEC imports magnet alloys used for its magnet production from China.

A Government Accountability Office (GAO) report illustrates the lack of U.S. presence in the REE global supply chain at each of the five stages of mining, separation, refining oxides into metal, fabrication of alloys, and the manufacturing of magnets and other components.[26] China produces 97% of the

REE raw materials, about 97% of rare earth oxides, and is the only exporter of commercial quantities of rare earth metals (Japan produces some metal for its own use for alloys and magnet production). About 90% of the metal alloys are produced in China (small production in the United States) and China manufactures 75% of the neodymium magnets and 60% of the samarium magnets. A small amount of samarium magnets are produced in the United States. Thus, even if rare earth production ramps up, much of the processing/alloying and metal fabrication would occur in China. According to investor analyst Jack Lifton, the rare earth metals are imported from China, then manufactured into military components in the United States or by an allied country.

Lifton states that many investors believe that for financing purposes, it is not enough to develop REE mining operations alone without building the value-added refining, metal production, and alloying capacity that would be needed to manufacture component parts for end-use products. According to Lifton, vertically integrated companies may be more desirable. It may be the only way to secure investor financing for REE production projects.[27] Joint ventures and consortiums could be formed to support production at various stages of the supply chain at optimal locations around the world. Each investor or producer could have equity and offtake commitments. Where U.S. firms and its allies invest is important in meeting the goal of providing a secure and stable supply of REEs, intermediate products, and component parts needed for the assembly of end-use products.

Role of China

State-run ("State-Key") labs in China have consistently been involved in research and development of REEs for over fifty years. There are two State-Key labs: (1) Rare Earth Materials Chemistry and Applications, which has focused on rare earth separation techniques and is affiliated with Peking University, and (2) Rare Earth Resource Utilization, which is associated with the Changchun Institute of Applied Chemistry. Additional labs concentrating on rare earth elements include the Baotou Research Institute of Rare Earths, the largest rare earth research institution in the world, established in 1963, and the General Research Institute for Nonferrous Metals established in 1952.[28] This long term outlook and investment has yielded significant results for China's rare earth industry.

Major iron deposits in the Bayan Obo in Inner Mongolia contain significant rare earth elements recovered as a byproduct or co-product of iron ore mining. China has pursued policies that would use Bayan Obo as the center of rare earth production and R&D. REEs are produced in the following provinces of China: Baotao (Mongolia) Shangling, Jiangxi, Guangdong, Hunan, Guangxi, Fujian, and Sischuan. Between 1978 and 1989, China's annual production of rare earth elements increased by 40%. Exports rose in the 1990s, driving down prices. In 2007, China had 130 neo-magnet producers with a total capacity of 80,000 tons. Output grew from 2,600 tons in 1996 to 39,000 tons in 2006.

Spurred by economic growth and increased consumer demand, China is ramping up for increased production of wind turbines, consumer electronics, and other sectors, which would require more of its domestic rare earth elements. Safety and environmental issues may eventually increase the costs of operations in China's rare earth industry as domestic consumption is becoming a priority for China. REE manufacturing is set to power China's surging demand for consumer electronics—cell phones, laptops and green energy technologies. According to the report by Hurst, China is anticipating going from 12 gigawatts (GW) of wind energy in 2009 to 100 GW in 2020. Neodymium magnets are needed for this growth.[29]

China's policy initiatives restrict the exports of rare earth raw materials, especially dysprosium, terbium, thulium, lutetium, yttrium, and the heavy rare earths. The export restrictions would not likely affect the downstream metal or magnets. According to Hurst, China wants an expanded and fully integrated REE industry where exports of value-added materials are preferred. It is common for a country to want to develop more value-added production and exports if it is possible. Hurst also suggests China wants to build strategic stockpiles of raw materials as South Korea and Japan have done, and thus have better control over global supply and prices.

RARE EARTH LEGISLATION IN THE 111TH CONGRESS

Congress is concerned about the potential problems with access to rare earth raw materials and downstream products used in many national security applications and clean energy technologies. Legislative proposals have been introduced in the 111th Congress to address these issues.

H.R. 4866, the Rare Earths Supply-Chain Technology and Resources Transformation Act of 2010

The House bill, **H.R. 4866** (Coffman) was introduced on March 29, 2010, has 16 co-sponsors, and was referred to the House Committee on Ways and Means, Subcommittee on Trade. The purpose of the bill is

> To reestablish a competitive domestic rare earths minerals production industry; a domestic rare earth processing, refining, purification, and metals production industry; a domestic rare earth metals alloying industry; and a domestic rare earth based magnet production industry and supply chain in the United States.

The bill would establish executive agents (at the Assistant Secretary level) from the Departments of Commerce, Defense, Energy, Interior, and State to serve on an interagency working group. The U.S. Trade Representative (USTR) and White House Office of Science and Technology Policy would also appoint representatives to the working group. The secretaries and the representatives appointed from above agencies would assess rare earth supply chain to determine which REEs are critical to national and economic security. Based on a critical designation, rare earth elements would be stockpiled in the national stockpile administered by the Defense Logistics Agency as part of the National Defense Stockpile. The DLA would, if necessary, make a commitment to purchase rare earth raw materials to process and refine raw materials including purchases made from China if necessary. Stockpiling would be terminated when working group agencies determine REEs are no longer critical to U.S. national security or economic well-being.

The USTR would review international trade practices of REE producers to examine possible market manipulation. Action before the WTO would be possible. Loan guarantees would be provided for supply chain development and the House bill would require a report that would describe "mechanisms" for obtaining loan guarantees for new supply chain development in the U.S. The Department of Defense (DOD) would issue guidance for obtaining loans for new defense supply chains and the Department of Energy (DOE) would issue guidance for development of a domestic supply chain for civilian and commercial purposes.

There is a Sense of Congress statement that would use the Defense Production Act (DPA) of 1950 to develop domestic rare earth supply chain and to provide workforce development and training to reestablish the United States as the preeminent center for rare earths production. R&D funding would also be authorized and if no projects are taking place under DPA, the Congress would want to be notified.

S. 3521, Rare Earths Supply Technology and Resources Transformation Act of 2010

A Senate proposal, **S. 3521**(Murkowski), similar to the House bill, would expedite permitting to increase domestic exploration and development of REEs. A Rare Earth Policy Task Force would be established and would be composed of the Secretaries of Interior, Energy, Defense, Commerce, State, and Agriculture. A representative from OMB, CEQ, and others as deemed necessary by the Secretary of the Interior would also serve on the task force. The Rare Earth Policy Task Force would monitor the acceleration of REE projects, review policies that discourage REE development, and report to Congress on results annually. The Secretaries of Interior, Energy, and others would assess supply chain vulnerability and determine elements "critical" to clean energy technologies. They would prepare a stockpile report to determine if rare earth materials critical for clean energy technology, national and economic security should be stockpiled and determine if legal authorities exist to procure stockpile. If a stockpile were established for clean energy, national defense, and economic security of the United States, it would contain rare earth oxides, and other storable forms of rare earths and alloys. DOE would issue guidance on obtaining loan guarantees to support a domestic supply chain for clean energy and defense technologies. The DPA would be used to reestablish a domestic supply chain of REEs. A sense of the Congress provision would fund workforce development and training and R&D.

H.R. 5136, the Fiscal Year 2011 National Defense Authorization Act

The House-passed bill (H.R. 5136) would require the Secretary of Defense to assess the rare earth material supply chain to determine if any of the materials were strategic or critical to national security. If the material is determined to be strategic, the Secretary would be required to develop a plan to ensure long-term availability by December 31, 2015. The Secretary shall submit a report to Congress on the assessment and the plan not later than 180 days after enactment of this legislation.

Also, based on congressional findings, among other things, there is an urgent need to eliminate U.S. vulnerability related to the supply of neodymium iron boron magnets and to restore the domestic capacity to manufacture sintered neodymium iron boron magnets used in defense applications. Within 90 days of enactment of this bill the Secretary of Defense would be required to submit a plan, to the appropriate congressional committees, to establish a domestic source of sintered neodymium iron boron magnets used in defense applications.

P.L. 111-84, the Fiscal Year 2010 National Defense Authorization Act

In the proposed House and Senate (H.R. 2647/S. 1390) versions of the defense authorization bill for 2010, Representative Mike Coffman and Senator Evan Bayh introduced legislation to direct the Comptroller General to determine the extent to which specific military weapons systems are currently dependent upon rare-earth materials and the degree to which the United States is dependent upon sources that could be interrupted or disrupted. The measure also directed DOD to describe the risks (both current and projected) involved in the United States' dependence on foreign sources of these materials, and any steps DOD has taken or plans to take to address any potential risks to national security.[30] The measure was passed in the Fiscal Year 2010 National Defense Authorization Act.[31]

POSSIBLE POLICY OPTIONS

Authorize and Appropriate Funding for a USGS Assessment

In addition to the two current legislative proposals, Congress could authorize and appropriate funding for a USGS comprehensive assessment to identify economically exploitable REE deposits (as a main product or co-product), and where REE could be exploited as a byproduct. Authorizations and additional appropriations could be supported for basic science research on substitutes and efforts at secondary recovery for REEs. Additionally, R&D may be necessary on how to proceed in the exploitation of high-thorium monazite deposits where REE could be produced as a byproduct.

Support and Encourage Greater Exploration for REE

Supporting/encouraging greater exploration for REE efforts in the United States, Australia, Africa, and Canada could be part of a broad international strategy. There are only a few companies in the world that can provide the exploration and development skills and technology for REE development. These few companies are located primarily in Canada, Australia, China,

South Africa, and the United States, and may form joint ventures or other types of alliances for R&D, and for exploration and development of REE deposits worldwide, including those in the United States. Whether there should be restrictions on these efforts in the United States is a question that Congress may ultimately choose to address.

Challenge China on Its Export Policy

Challenging China on its export restrictions through the WTO would involve filing a dispute based on WTO rules that generally prohibit members from imposing restrictions (i.e., quotas) or other restraints (e.g., minimum prices or licensing) on exports. In June 2009, the United States filed a dispute over raw material exports from China, which included: bauxite, coke, fluorspar, magnesium, manganese, silicon carbide, silicon metal, yellow phosphorus and zinc.[32] Some REE analysts assert that China sets export restrictions to meet growing Chinese demand for raw materials and to force the manufacturing of end-use products in China.[33]

Establish a Stockpile

Establishing a government-run economic stockpile and/or private-sector stockpiles that would contain supplies of specific REE broadly needed for "green initiatives" and defense applications is a policy advocated by some in industry and government. This may be a prudent investment. Generally, stockpiles and stockpile releases could have an impact on prices and supply but would also ensure supplies of REE materials (oxides, metals, etc.) during times of normal supply bottlenecks. An economic stockpile could be costly and risky, as prices and technology may change the composition of REEs that are needed in the economy.

According to USGS,[34] DOD along with USGS is examining which of the REEs might be necessary in the National Defense Stockpile (NDS). In the recent past, NDS materials were stored for wartime use based on a three-year war scenario. Some of the rare earth elements contained in the National Defense Stockpile were sold off by 1998. However, rare earth elements were never classified as strategic minerals.[35] DOD had stockpiled some yttrium but has since sold it off, and none of the REEs have been classified as strategic materials. A critical question for stockpile development would be: What materials along the supply chain should be stockpiled? For example, should the stockpile contain rare earth oxides or alloyed magnets which contain the REEs, or some combination of products?

The National Research Council (NRC) has produced an in-depth report on minerals critical to the U.S. economy and offers its analysis as described here: "... most critical minerals are both essential in use (difficult to substitute for) and prone to supply restrictions."[36] While the NRC report is based on several availability criteria used to rank minerals for criticality (geological, technical, environmental and social, political, and economic), REEs were determined to be critical materials assessed at a high supply risk and the possibility of severe impacts if supplies were restricted. Some of the REE applications are viewed as more important than others and some are at greater risk than others, namely the Heavy Rare Earth Elements (HREEs), as substitutes are unavailable or not as effective.[37]

The federal government and private sectors are beginning to address how to secure reliable rare earth materials (raw materials through metals and alloys) from China and non-Chinese sources in the short term, and how to rebuild the U.S. supply chain for the long term.

End Notes

[1] U. S. mineral policies provide a framework for the development of domestic metal mineral resources and for securing supplies from foreign sources. Specifically, the Mining and Minerals Policy Act of 1970 (30 U.S.C. §21a) declared that it is in the national interest of the United States to foster the development of the domestic mining industry "... including the use

of recycling and scrap." The National Materials and Minerals Policy, Research and Development Act of 1980 (30 U.S.C. 1601), among other things, declares that it is the continuing policy of the United States to promote an adequate and stable supply of materials necessary to maintain national security, economic well-being and industrial production, with appropriate attention to a long-term balance between resource production, energy use, a healthy environment, natural resources conservation, and social needs.

[2] U.S. Department of the Interior (DOI), Geological Survey (USGS), *Minerals Yearbook, Volume 1*, 2007, Rare Earths (Advance Release).

[3] DOI/USGS, *Rare Earth Elements—Critical Resources for High Technology*, Fact Sheet 087-02.

[4] Rare Earth Minerals: The Indispensable Resource for Clean Energy Technologies, presented at Minerals for a Green Society Conference, Washington, DC, by Mark A. Smith, CEO, Molycorp Minerals, February 4, 2010.

[5] DOI/USGS Minerals Yearbook, Volume 1, 2007.

[6] This section was prepared by Valerie Grasso, CRS Foreign Affairs, Defense, and Trade Division.

[7] The following information was accessed on March 10, 2010, from the Defense National Stockpile Center website, at https://www.dnsc.dla.mil/inside.asp., and GlobalSecurity.org, at http://www.globalsecurity.org/military dnsc.htm. "The National Stockpile operates under authority of the Strategic and Critical Materials Stockpiling Act (50 U.S.C. 98-h-2(a)). This act provides that strategic and critical materials are stockpiled in the interest of national defense to preclude a dangerous and costly dependence upon foreign sources of supply in times of national emergency. The Defense National Stockpile Center administers the storage, management, and disposal of the Nation's inventory of strategic and critical materials essential to the military and industrial requirements of the United States in times of national emergency. The Congress of the United States has authorized the Defense National Stockpile Center to sell commodities that are excess to Department of Defense needs. Since 1993, DNSC sales have totaled approximately $6.6 billion."

[8] For further discussion on the Berry Amendment and the Specialty Metals provision, see CRS Report RL3 1236, *The Berry Amendment: Requiring Defense Procurement to Come from Domestic Sources*, and CRS Report RL33751, *The Specialty Metal Provision and the Berry Amendment: Issues for Congress*, both by Valerie Bailey Grasso.

[9] For further discussion on precision-guided munitions, see CRS Report RL33539, *Intelligence Issues for Congress,* by Richard A. Best Jr., and CRS Report RL30727, *Airborne Intelligence, Surveillance, and Reconnaissance (ISR): The U2 Aircraft and Global Hawk UAV Programs,* by Richard A. Best Jr. and Christopher Bolkcom.

[10] Haxel, Gordon B, Hendricks, James B., and Oris, Greta J. Rare Earth Elements: Critical Resources for High Technology. *U.S. Geological Survey*, Fact Sheet 087-82, accessed online on March 10, 2010, at http://pubs.usgs.gov/fs/ 2002/fs087-02/.

[11] Green Jeffrey A. Defense, Energy Markets Should Brace for Shortages of Key Materials. *National Defense Industrial Association*, October 2009; U.S. Lacks Data on Supply of Minerals Critical to Economy, National Security; Defense Stockpile is Ineffective. National Academy of Sciences, October 5, 2007, accessed online at http://www8.nationalacademies.org/onpinews/newsitem.aspx?RecordID=10052007.

[12] Chang, Gordon G. Why Does China Want BP? *Forbes.com*, July 14, 2010. http://www.forbes.com/2010/07/14/ bpchina-oil

[13] Magnet Materials Supply Chain Players Propose Six-Point Plan to Address Impending Rare Earths Crisis, *USMMA*, February 4, 2010, accessed online at http://www.usmagnetmaterials. com/?p=74.

[14] Ibid.

[15] Bastnaesite is mineral with the formula (Ce, La)CO3(F,OH) that may contain other rare earth elements.

[16] Monazite is a mineral with the formula (Ce, La, Nd, Th)PO4 that may contain other rare earth elements.

[17] DOI/USGS Circular 930 N, *International Strategic Minerals Inventory Summary Report—Rare Earth Oxides*, by Wayne Jackson and Grey Christiansen, 1993.

[18] Dr. Anthony N. Mariano, *The Nature of Economic REE and Y Minerals on a World Level*, presented at the *MIT Energy Initiative Worksho*p, April 29, 2010.

[19] Ibid.

[20] The Jack Lifton Report, *The Rare Earth Crisis—Part I*, by Jack Lifton, October 2009.

[21] *Rare Earth Strategic Supplies More Important Than Price, Industrial Metals/Minerals Interview of Jack Lifton* by The Gold Report, December 14, 2009.

[22] Ibid.

[23] Ibid.

[24] "China Cuts Rare Earth Export Quota," Bloomberg News Report, July 9, 2010.

[25] Jack Lifton, "Is The Rare Earth Supply Crisis Due to Peak Production Capability or Capacity," michaelperelman.worldpress.com, September 6, 2009.

[26] U.S. Government Accountability Office, *Rare Earth Materials in the Defense Supply Chain*, GAO-10-617R, April 1, 2010.

[27] Op. Cit., Lifton Interview by The Gold Report, December 14, 2009.

[28] China's Rare Earth Elements Industry: What Can the West Learn? by Cindy Hurst, Institute for the Analysis of Global Security, March 2010.

[29] Ibid.

[30] McCormack, Richard. China's Complete Control of Global High-Tech Magnet Industry Raises U.S. National Security Alarms. *Manufacturing & Technology News*, Vol. 16, No. 16, September 30, 2009, viewed online at http://www.manufacturingnews.com/news/09/0930/magnets

[31] P.L. 111-84 was signed into law on October 28, 2009. The text, as it appears in the bill, reads as follows.

Sec. 843. Report on Rare Earth Materials in the Defense Supply Chain.

(a) Report Required.—Not later than April 1, 2010, the Comptroller General shall submit to the Committees on Armed Services of the Senate and House of Representatives a report on rare earth materials in the supply chain of the Department of Defense.

(b) Matters Addressed.—The report required by subsection (a) shall address, at a minimum, the following: (1) An analysis of the current and projected domestic and worldwide availability of rare earths for use in defense systems, including an analysis of projected availability of these materials in the export market. (2) An analysis of actions or events outside the control of the Government of the United States that could restrict the access of the Department of Defense to rare earth materials, such as past procurements and attempted procurements of rare earth mines and mineral rights. (3) A determination as to which defense systems are currently dependent on, or projected to become dependent on, rare earth materials, particularly neodymium iron boron magnets, whose supply could be restricted—(A) by actions or events identified pursuant to paragraph (2); or (B) by other actions or events outside the control of the Government of the United States. (4) The risk to national security, if any, of the dependencies (current or projected) identified pursuant to paragraph (3). (5) Any steps that the Department of Defense has taken or is planning to take to address any such risk to national security. (6) Such recommendations for further action to address the matters covered by the report as the Comptroller General considers appropriate. (c) Definitions.—In this section: (1) The term "rare earth" means the chemical elements, all metals, beginning with lanthanum, atomic number 57, and including all of the natural chemical elements in the periodic table following lanthanum up to and including lutetium, element number 71. The term also includes the elements yttrium and scandium. (2) The term "rare earth material" includes rare earth ores, semi-finished rare earth products, and components containing rare earth materials.

[32] Office of the U.S. Trade Representative, Press Release, "WTO Case Challenging China's Export Restraints on Raw Material Inputs," June 23, 2009.

[33] Irma Venter, "Investors take closer look at rare earth elements as technology, green revolution pick up pace," Mining Weekly Online, http://www.miningweekly.com, September 18, 2009.

[34] Phone interview with Jim Hedrick, Rare Earth Specialist, USGS, October 1, 2009.

[35] For a discussion of the strategic materials for defense uses, see CRS Report RL33751, *The Specialty Metal Provision and the Berry Amendment: Issues for Congress*, by Valerie Bailey Grasso.

[36] National Research Council, *Minerals, Critical Minerals, and the U.S. Economy*, National Academies Press, 2008.

[37] DOI/USGS, Minerals Yearbook, Volume 1, 2007.

In: Rare Earth Minerals: Policies and Issues   ISBN: 978-1-61122-310-1
Editor: Steven M. Franks   © 2011Nova Science Publishers, Inc.

Chapter 2

RARE EARTH MATERIALS IN THE DEFENSE SUPPLY CHAIN

United States Government Accountability Office

April 14, 2010

The Honorable Carl Levin
Chairman
The Honorable John McCain
Ranking Member
Committee on Armed Services
United States Senate

The Honorable Ike Skelton
Chairman
The Honorable Howard P. "Buck" McKeon
Ranking Member
Committee on Armed Services
House of Representatives

Subject: *Rare Earth Materials in the Defense Supply Chain*

This letter formally transmits the enclosed briefing in response to the National Defense Authorization Act for Fiscal Year 2010 (Pub. L. No. 111-84), which required GAO to submit a report on rare earth materials in the defense supply chain to the Committees on Armed Services of the Senate and House of Representatives by April 1, 2010. As required, we provided a copy of this briefing to the committees on April 1, 2010, and subsequently briefed the Senate Armed Services Committee staff on April 5, 2010, and the House Armed Services Committee staff on April 6, 2010.

We are sending copies of this report to the appropriate congressional committees. We are also sending copies to the Secretaries of Defense, Commerce, Energy, and the Interior. This report is also available at no charge on the GAO Web site at http://www.gao.gov. Should you or your staff have any questions concerning this report, please contact me at (202) 512-4841 or martinb@gao.gov. Contact points for our Offices of Congressional Relations and Public Affairs may be found on the last page of this report. Key contributors to this report were John Neumann, Assistant Director; James Kim; Erin Carson; Brent Corby; Marie Ahearn; Barbara El Osta; and Morgan Delaney Ramaker.

Belva M. Martin
Acting Director
Acquisition and Sourcing Management Enclosure

CONTENTS

INTRODUCTION

- Rare earth elements are used in many applications for their magnetic and other unique properties. These include the 17 chemical elements beginning with lanthanum, element number 57 in the periodic table, up to and including lutetium, element number 71, as well as yttrium and scandium, which have similar properties.
- Rare earth materials—rare earth ores, oxides, metals, alloys, semifinished rare earth products, and components containing rare earth materials—are used in a variety of commercial and military applications, such as cell phones, computer hard drives, and Department of Defense (DOD) precision-guided munitions. Some of these applications rely on permanent rare earth magnets that have unique properties, such as the ability to withstand demagnetization at very high temperatures.
- Media reports have noted worldwide availability of these materials may be limited to a few overseas sources—primarily China.

OBJECTIVES, SCOPE, AND METHODOLOGY

- The National Defense Authorization Act for Fiscal Year 2010, Section 843, directed GAO to submit a report on rare earth materials in the DOD supply chain.[1]
- Objectives:
  What does existing information show about current sources and projected availability of rare earth materials?
  Which defense systems have been identified as dependent on rare earth materials?
  What national security risks has DOD identified due to rare earth material dependencies, and what actions has it taken?
- To conduct our work, we obtained documentation and interviewed officials to determine the current sources and projected availability of rare earth materials and national security risks DOD has identified and actions DOD has taken.
- We contacted federal agencies and offices, including the following: Department of the Interior,

- U.S. Geological Survey (USGS);
  Department of Commerce,
  - Bureau of Industry and Security,
  - International Trade Administration;
  Department of Energy,
  - Vehicle Technology Program,
  - Wind Technologies Program,
  - Energy Information Administration, and
  - Ames (Iowa) Laboratory;
- DOD,
  - Office of the Secretary of Defense – Industrial Policy, Office of Technology Transition, Defense Research and Engineering, Science and Technology, and Net Assessment,
  - Defense Logistics Agency, Defense Contract Management Agency, Defense Intelligence Agency, Defense Advanced Research Projects Agency, and Missile Defense Agency,
  - Military departments including: Army Research, Development and Engineering; Army Acquisition, Logistics, and Technology; Army Tank and Automotive Command; Air Force Research Lab Materials and Manufacturing; Naval Surface Warfare Center; Naval Research Laboratory; Navy Research, Development, and Acquisition; and Navy Program Executive Office for Ships.
- We contacted members of industry and academia, including the following:
  - Institute for Defense Analyses, a nonprofit corporation that administers federally funded research and development centers;
  - Academic experts at the University of Delaware and Northeastern University;[2]
  - The National Academies;
  - Rare Earth Industry and Technology Association; and
  - Selected rare earth suppliers from each stage of the supply chain.[3]
- To determine which defense systems are currently dependent on, or projected to become dependent on, rare earth materials, we held discussions with and gathered evidence from government, industry, and academic officials, who identified certain defense systems that use and will continue to use rare earth materials.[4] In addition, we analyzed the supply chains of two specific defense systems to provide illustrative examples of systems that use rare earth materials.

- We used industry reports and data to evaluate the projected worldwide supply and demand of rare earth materials. Uncertainty exists in these estimates due to the assumptions made by different projections. As our findings do not rely on precise estimates of the amount of rare earth material available throughout the world, we found these data to be sufficiently reliable for the purposes of our reporting.
- We conducted this performance audit from January 2010 through April 2010 in accordance with generally accepted government auditing standards. Those standards require that we plan and perform the audit to obtain sufficient, appropriate evidence to provide a reasonable basis for our findings and conclusions based on our audit objectives. We believe that the evidence obtained provides a reasonable basis for our findings and conclusions based on our audit objectives.

BACKGROUND: THE RARE EARTH ELEMENTS

- The term "rare earth" denotes the group of 17 chemically similar metallic elements, including lanthanum, cerium, praseodymium, neodymium, promethium, samarium, europium, gadolinium, terbium, dysprosium, holmium, erbium, thulium, ytterbium, lutetium, scandium, and yttrium.

Figure 1: Rare Earth Elements in the Periodic Table

Source: GAO graphic based on USGS data.

Table 1: Examples of Rare Earth Elements Used in Commercial Products

Rare Earth Element Used	Commercial Product
Neodymium, praseodymium, dysprosium, terbium, lanthanum, cerium	Hybrid electric motors and hybrid batteries
Neodymium, praseodymium, terbium, dysprosium	Computer hard drives, mobile phones, and cameras
Promethium	Portable x-ray units
Scandium	Stadium lights
Europium, yttrium, terbium, lanthanum	Energy-efficient light bulbs
Europium, yttrium	Fiber optics
Cerium, lanthanum, neodymium, europium	Glass additives

Source: GAO analysis of government and industry data.

- Rare earths are often classified into two groups: Heavy Rare Earth (HREE), and Light Rare Earth (LREE), according to their atomic weights and location on the periodic table.

Rare Earth Materials Are Used in Multiple Commercial Products

- Rare earth elements are used in materials for a number of commercial products, including hybrid cars, wind power turbines, computer hard drives, and cell phones.

Rare Earth Material Production Requires a Number of Key Processing Steps

- Rare earth materials require a number of processing stages before they can be used in an application:
  mining rare earth ore from the mineral deposit;
  separating the rare earth ore into individual rare earth oxides;
  refining the rare earth oxides into metals with different purity levels;[5]
  forming the metals into rare earth alloys; and
  manufacturing the alloys into components, such as permanent magnets, used in defense and commercial applications.

DOD Responsibilities for Managing Supplier Base

- DOD's Office of the Director of Industrial Policy sustains an environment that ensures the industrial base on which DOD depends is reliable, cost-effective, and sufficient to meet DOD requirements. It routinely identifies and works to mitigate short-term supplier-base gaps when these gaps span multiple DOD components.
- The Defense National Stockpile maintains and manages strategic and critical materials.
- DOD military service components (Army, Navy, and Air Force) assess supplier-base issues for existing defense programs or sectors.

SUMMARY

Objective 1: Current and Projected Availability

- While rare earth ore deposits are geographically diverse, current capabilities to process rare earth metals into finished materials are limited mostly to Chinese sources.
- The United States previously performed all stages of the rare earth material supply chain, but now most rare earth materials processing is performed in China, giving it a dominant position that could affect worldwide supply and prices.
- Based on industry estimates, rebuilding a U.S. rare earth supply chain may take up to 15 years and is dependent on several factors, including securing capital investments in processing infrastructure, developing new technologies, and acquiring patents, which are currently held by international companies.

Objective 2: Defense System Dependency

- DOD is in the early stages of assessing its dependency on rare earth materials and is planning to complete its study by the end of September 2010.

- Government and industry officials have identified a wide variety of defense systems and components that are dependent on rare earth materials for functionality and are provided by lower-tier subcontractors in the supply chain.
- Defense systems will likely continue to depend on rare earth materials, based on their life cycles and lack of effective substitutes.
- We found examples of components in defense systems that use Chinese sources for rare earth materials and are provided by lower- tier subcontractors.

Objective 3: DOD Identified Risks and Actions Taken

- DOD has not yet identified national security risks or taken departmentwide action to address rare earth material dependency, but expects to consider these issues in its ongoing study expected to be completed by the end of September 2010.
- Some DOD components, other federal agencies, and companies are taking initial steps to limit their reliance on rare earth materials or expand the existing supplier base.

OBJECTIVE 1: RARE EARTH ORE DEPOSITS ARE GEOGRAPHICALLY DIVERSE

- Significant rare earth ore reserves exist in China as well as other worldwide locations, including the United States.

- The less-abundant, and more-valuable, heavy rare earth ore deposits are currently found in southern China, but such deposits have also been identified in Australia, Greenland, Canada, and the United States.
- According to industry, rare earth deposits in the United States, Canada, Australia, and South Africa could be mined by 2014.

Table 2: World Mine Reserves and Production

Country	Reserves (t REO)[a]	2009 Mine Production (t REO)
United States	13,000,000	0
Australia	5,400,000	0
Brazil	48,000	650
China	36,000,000	120,000
Commonwealth of Independent States (CIS)[b]	19,000,000	N/A[c]
India	3,100,000	2,700
Malaysia	30,000	380
Other Countries	22,000,000	N/A
World Total (rounded)	**99,000,000**	**124,000**

Source: USGS.

Note: Data are from the Mineral Commodity Summaries 2010.

[a]According to USGS, reserves are the part of the reserve base that could be economically extracted or produced at the time of determination but need not signify that extraction facilities are in place and operative. t REO = metric tons of rare earth oxide.

[b]Regional association made up of former Soviet republics.

Figure 2: History of the U.S. Rare Earth Industry

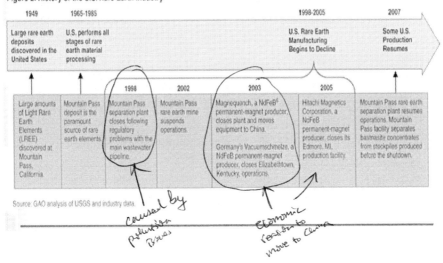

Source: GAO analysis of USGS and industry data.

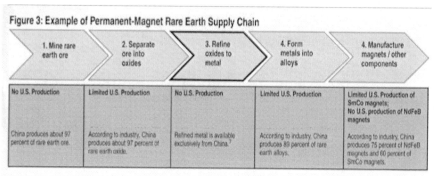

Figure 3: Example of Permanent-Magnet Rare Earth Supply Chain

1. Mine rare earth ore	2. Separate ore into oxides	3. Refine oxides to metal	4. Form metals into alloys	4. Manufacture magnets / other components
No U.S. Production	Limited U.S. Production	No U.S. Production	Limited U.S. Production	Limited U.S. Production of SmCo magnets; No U.S. production of NdFeB magnets
China produces about 97 percent of rare earth ore.	According to industry, China produces about 97 percent of rare earth oxide.	Refined metal is available exclusively from China.[7]	According to industry, China produces 89 percent of rare earth alloys.	According to industry, China produces 75 percent of NdFeB magnets and 60 percent of SmCo magnets.

Source: GAO analysis of industry data.

U.S. Industry Previously Performed All Stages of the Supply Chain

- U.S. industry previously performed all stages of the rare earth material supply chain, and the Mountain Pass mine in California produced the majority of the global supply of rare earth materials.

Most Rare Earth Material Processing Occurs in China

- Most rare earth material processing now occurs in China. In 2009, China produced about 97 percent of rare earth oxides.
- Officials of the minerals and rare earth company that owns the Mountain Pass mine expect that by 2012 it will achieve full-scale production of mining and separating cerium, lanthanum, praseodymium, and neodymium oxides.
- The Mountain Pass facility does not currently have the full capability needed to refine the oxides into pure rare earth metals.
- According to industry data, refined rare earth metals are almost exclusively available from China.
- The United States has the expertise but lacks the manufacturing assets and facilities to refine oxides to metals.
- The United States is not currently producing neodymium iron boron (NeFeB) permanent magnets and has only one samarium cobalt (SmCo) magnet producer.

China's Market Dominance May Affect Future U.S. Availability of Rare Earth Materials

- According to government and industry data, the future availability of materials from some rare earth elements—including neodymium, dysprosium, and terbium—is largely controlled by Chinese suppliers.
- China's dominant position in the rare earths market gives it market power,[8] which could affect global rare earth supply and prices. In addition:

  China has adopted domestic production quotas on rare earth materials and decreased its export quotas, which increases prices in the Chinese and world rare earth materials markets.

  China increased export taxes on all rare earth materials to a range of 15 to 25 percent, which increases the price of inputs for non-Chinese competitors.
- Some government and rare earth industry officials believe that China plans on greater vertical integration of the rare earth materials market in the future, which would increase China's total market power and dominance.
- While China is currently exporting rare earth oxides and metals, some rare earth industry officials believe that in the future China will only export finished rare earth material products with higher value.

↑↑↑ What are they? How are they made?

Rebuilding a U.S. Supply Chain Is Dependent on Several Factors

- Although the Mountain Pass mine is the largest non-Chinese rare earth deposit in the world, the mine currently lacks the manufacturing assets and facilities to process the rare earth ore into finished components, such as permanent magnets.

JV. opport btw US + China

- The Mountain Pass mine also does not have substantial amounts of heavy rare earth elements, such as dysprosium, which provide much of the heat-resistant qualities of permanent magnets used in many industry and defense applications.
- Other U.S. rare earth deposits exist, such as those in Idaho, Montana, Colorado, Missouri, Utah, and Wyoming, but these deposits are still in early exploratory stages of development. Once a company has secured the necessary capital to start a mine, government and industry officials

said it can take from 7 to 15 years to bring a property fully online, largely due to the time it takes to comply with multiple state and federal regulations.

- Other factors may affect the rebuilding of a U.S. supply chain:

Capital investment—Industry officials noted that processing companies will need to secure a large amount of capital to begin operations, but investors are concerned about the possibility of the Chinese undercutting U.S. prices and negatively affecting their return on investments.

Processing plants—Industry officials said it would take from 2 to 5 years to develop a pilot plant that could refine oxides to metal using new technologies, and companies with existing infrastructure said they would not restart metal production without a consistent source of oxides outside of China.

Environmental concerns—Some rare earth minerals are accompanied by radioactive products, such as thorium and radium, which make extraction difficult and costly. In addition, U.S. mines and processing facilities must comply with environmental regulations.

New technologies—Some academic experts believe that new processing technologies are needed in order to compete with Chinese producers on price, and academic experts do not believe these technologies will be available on a full production scale for up to 4 years and will require large start-up costs.

Intellectual property rights—Japanese and other foreign companies currently own key technology patents for manufacturing neodymium iron boron magnets. Some of these patents do not expire until 2014. As a result, companies preparing to enter the neodymium iron boron magnet market in the United States must wait for the patents to expire.

The development of alternatives to rare earth materials could reduce the demand and dependence on rare earth materials in 10 to 15 years, but these materials might not meet current application requirements.

OBJECTIVE 2: DOD HAS BEGUN ASSESSING RARE EARTH MATERIAL DEPENDENCY

- DOD has begun a review, on its own initiative, assessing its dependency on rare earth materials, that it plans to complete by the end of September 2010. DOD plans to assess its use of these materials as well as vulnerabilities in the supply chain.9
- In 2008, DOD Industrial Policy conducted an initial inquiry of DOD departments and agencies to identify strategic and critical materials required for national defense purposes. Although respondents identified a range of systems and components whose production could potentially be delayed due to a lack of availability of rare earth materials, DOD officials stated that this information was not based on a formal study on the use of rare earth materials in these systems.

Rare Earth Materials Are Widely Used and Lack Substitutes

- According to government, industry, and academic officials, the use of rare earth materials is widespread in defense systems. These include, among others,
  precision-guided munitions,
  lasers,
  communication systems,
  radar systems,
  avionics,
  night vision equipment, and
  satellites.
- Officials emphasized the significance of the widespread use of commercial-off-the-shelf products in defense systems that include rare earth materials, such as computer hard drives.
- Officials also cited specific components within defense systems that rely on rare earth materials, such as traveling-wave tubes, which amplify radio-frequency signals using rare earth permanent magnets.
- Government and industry officials told us that where rare earth materials are used in defense systems, the materials are responsible for the functionality of the component and would be difficult to replace

without losing performance. For example, fin actuators used in precision-guided munitions are specifically designed around the capabilities of neodymium iron boron rare earth magnets.

Defense Systems Will Continue to Rely on Rare Earth Materials

- Many defense systems will continue to use rare earth materials in the future based on their life cycles and the lack of effective substitutes.
  For example, the Aegis Spy-1 radar, which is expected to be used for 35 years, has samarium cobalt magnet components that will need to be replaced during the radar's lifetime.
  According to officials, defense system components that have rare earth materials in them will wear out and need to be replaced.
  Defense officials said that future generations of some defense system components, such as transmit and receive modules for radars, will continue to depend on rare earth materials. Moreover, in some cases, new systems in development will also rely on components that depend on rare earth materials.

DOD Defense Systems Use Rare Earth Materials from China

- GAO analysis shows that subcontractors at the lower tiers of the supply chain use rare earth materials sourced from China to produce components used in larger defense systems.
- For example, the DDG-51 Hybrid Electric Drive Ship Program uses permanent-magnet motors using neodymium magnets from China.
- For example, the M1A2 Abrams tank has a reference and navigation system that uses samarium cobalt (SmCo) permanent magnets. The samarium metal used in these magnets comes from China.

Figure 4: Neodymium iron boron (NeFeB) magnets used in DDG-51

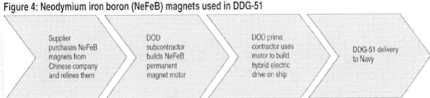

Figure 5: Samarium used in M1A2 Abrams tank

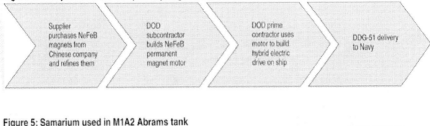

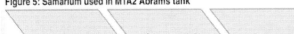

Source: GAO analysis of data from government, defense contractors, and rare earth material suppliers.

OBJECTIVE 3: DOD IN PROCESS OF IDENTIFYING DEPARTMENTWIDE SECURITY RISKS

- DOD has not yet identified departmentwide national security risks due to rare earth material dependencies and is in the process of assessing such risks.

    While Industrial Policy is aware of rare earth material supply concerns raised by industry and in its initial 2008 inquiry, officials also noted that as part of the office's current study, to be completed by the end of September 2010, they will address vulnerabilities in the supply chain and include recommendations to mitigate any potential risks of supply disruption.

    DOD has also been involved in efforts to transform the National Defense Stockpile so that materials not produced domestically will be available to support defense needs.

    A 2009 National Defense Stockpile configuration report identified lanthanum, cerium, europium, and gadolinium as having already caused some kind of weapon system production delay and recommended further study to determine the severity of the delays.[10]

- Industrial Policy has existing criteria in the *Defense Acquisition Guidebook* for when program offices should elevate supplier base concerns. These are when an item is produced by a single or sole-source supplier and meets one or more of the following criteria: (1) is used by three or more programs; (2) represents an obsolete, emerging, or enabling technology; (3) requires 12 months or more to manufacture; or (4) has limited surge production capability.
- Generally, Industrial Policy can help DOD offices address a supplier gap or vulnerability when requested. For example, while not related to rare earth materials, Industrial Policy worked with the Army to request a waiver that would allow the Hellfire Missile program to procure a chemical from China that was no longer produced in the United States. This allowed the program to explore a longer-term solution to develop a domestic source for the chemical.

Some DOD Components Have Taken Steps to Address Rare Earth Risks

- Apart from Industrial Policy's current study, DOD components are also taking steps to address rare earth risks.

  Air Force's Materials and Manufacturing Directorate examined the availability of rare earth materials and manufacturers of rare earth magnets in a 2003 internal report, which raised concerns about U.S. dependency on Chinese rare earth materials and U.S. industry's lack of intellectual property rights to produce neodymium iron boron magnets. An Air Force industrial base official told us that he was unaware of any actions taken to address the issues raised by the report. However, as we note in this briefing, DOD is in the process of studying these issues.

  Army's Armament Research Center and the Naval Surface Warfare Center have begun informal efforts to understand the extent of their current and future dependencies on rare earth materials.

  Also, in 2006, Navy considered funding the Mountain Pass mine and processing facility under a Title III[11] program to secure a domestic source of supply for rare earth materials but ultimately did not award a contract for that purpose as it lost interest in the project.

Although DOD has initiated a Title III program for domestic production of traveling-wave tubes, the program does not address domestic sources for the rare earth materials that are required for their production.

Other Government Agencies and Industry also Starting to Address Rare Earth Risks

- Several government agencies have made efforts, in which DOD participated, to address rare earth risks.
  - The Department of Commerce assembled a roundtable to review governmentwide options in addressing potential rare earth shortages.
  - The Office of Science and Technology Policy in the Executive Office of the President recently hosted an interagency meeting to discuss rare earth materials supply and demand and plans ongoing interagency coordination on the issue.
- The Department of Energy reported that it has several research and development efforts to develop non-rare-earth material-dependent motors, reduced rare earth material usage in magnets, and alternatives to rare earth dependent wind generators. In addition, the department recently announced that it will develop a strategic plan for addressing the role of rare earth and other strategic materials in clean energy technologies.
- A major defense contractor is informally surveying its suppliers to understand rare earth materials use in its defense system components and determine alternative solutions to their use.
  Rare earth industry and defense contractors have raised concerns about the Chinese monopoly for rare earth metals.

AGENCY COMMENTS

- We provided a draft of this briefing to DOD and the Departments of Commerce, Energy, and the Interior. DOD, Commerce, and Interior provided technical comments, which we incorporated as appropriate. Energy provided no comments.

End Notes

[1] Pub. L. No. 111-84 (2009).

[2] We selected a nongeneralizable sample of academics recommended to us through interviews.

[3] We selected a nongeneralizable sample of suppliers representing each processing step from mining to end product based on interviews with government and industry officials.

[4] We contacted three of the top five defense contractors, as identified by DOD based on contract award value for fiscal year 2009, as well as selected subcontractors identified by government and industry officials as producers of components containing rare earth materials. These contractors are not intended to be representative of the entire defense supplier base.

[5] Metallurgists refer to conversion of oxides into metals as reduction. For the purposes of this briefing, we refer to this step as refining.

[6] Neodymium iron boron.

[7] According to industry, only Chinese companies are producing and selling commercial quantities of rare earth metals. While some Japanese companies produce rare earth metals in a limited capacity, they do not offer these metals as a product but use them to produce alloys and magnets and are dependent on China for rare earth ore. One company in the United Kingdom produces a small quantity of samarium cobalt metal, but also relies on oxides and metals from China.

[8] Market power is defined as the ability of sellers to exert influence over the price or quantity of a good, service, or commodity exchanged in a market.

[9] USGS will conduct a portion of the study that focuses on rare earth element reserves and resources. The Defense Contract Management Agency's Industrial Analysis Center will review trends in pricing of rare earth materials and assess domestic rare earth material production capability.

[10] Industrial Policy noted that the stockpile report relied on the same data collected by DOD's 2008 inquiry, which indicated that only one DOD office reported actual production delays due to rare earth material shortages.

[11] Title III of the Defense Production Act of 1950, as amended, provides financial incentives to domestic firms to invest in production capabilities for national defense needs. 50 U.S.C. App. §§ 2091 et seq.

In: Rare Earth Minerals: Policies and Issues   ISBN: 978-1-61122-310-1
Editor: Steven M. Franks   © 2011 Nova Science Publishers, Inc.

Chapter 3

OPENING STATEMENT OF CHAIRMAN BART GORDON, BEFORE THE COMMITTEE ON SCIENCE AND TECHNOLOGY, HEARING ON "RARE EARTH MINERALS AND 21ST CENTURY INDUSTRY"

Bart Gordon

I'd like to thank Chairman Miller for calling this hearing. Last September, I saw an article on this issue that raised a number of questions in my mind about whether the Committee and the Congress were doing enough to support American business and American jobs.

Rare earths are an essential component in a wide array of emerging industries.

This is not the first time the Committee has been concerned with the competitive implications of materials such as rare earths. In 1980—30 years ago—this Committee established a national minerals and materials policy. One core element in that legislation was the call to support for "a vigorous, comprehensive and coordinated program of materials research and development."

Unfortunately, over successive administrations, the effort to keep that program going fell apart. Now, it is time to ask whether we need to revive a coordinated effort to level the playing field in rare earths.

In particular, I want to learn if there is a need for increased research and development to help address this Nation's rare earth shortage, or if we need to re-orient the research we already have underway.

Based on my review of the written submissions, it appears that we could benefit from more research both in basic and applied materials sciences.

Rare earths are not the only materials in which the U.S. is largely or exclusively dependent on foreign sources. According to the U.S. Geological Survey, there are eighteen other minerals and materials where the United States is completely dependent on foreign sources.

Someone needs to be telling us what's going on with those before we read about it in the New York Times. Legislation may be the best way to institutionalize a renewed focus and expanded commitment to identifying shortages and needs before they become a crisis.

Again, Mr. Chairman, I appreciate you holding this hearing and expect a stimulating discussion. I yield back my time.

In: Rare Earth Minerals: Policies and Issues   ISBN: 978-1-61122-310-1
Editor: Steven M. Franks   © 2011 Nova Science Publishers, Inc.

Chapter 4

OPENING STATEMENT OF CHAIRMAN BRAD MILLER, BEFORE THE COMMITTEE ON SCIENCE AND TECHNOLOGY, HEARING IN "RARE EARTH MINERALS AND 21ST CENTURY INDUSTRY"

Brad Miller

Welcome to our hearing this afternoon on something most of us have never heard of at all, or promptly forgot after our test on the Periodic Table in high school chemistry. Today we will be discussing rare earth elements, which aren't really all that rare. Rare earth elements are crucial to making the magnets and batteries needed for the energy industry of the 21st Century. With a little of one of these elements you can get a smaller, more powerful magnet, or an aircraft engine that operates at higher temperatures or a fiber-optic cable that can carry your phone call much greater distances.

The United States, not so long ago, was the world leader in producing and exporting rare earths. Today, China is the world's leader. We're having this hearing in part to recognize that the Chinese have some different ideas about how to get the greatest benefit from this suddenly-valuable commodity beyond simply digging it up and selling it to those who want to use it in their high-tech manufacturing. China appears to view rare earths as one of the incentives they can offer a technology firm scouting for a new plant location. How do we compete in attracting and retaining manufacturing firms that need access to rare earth elements in light of China's current near monopoly, and their willingness to use their monopoly power to our disadvantage?

The most immediate step would be to get some competition back into the supply of rare earths. One of our witnesses, Mr. Mark Smith, is proposing to do just that. His company owns a mine that could produce many rare earth elements if it were to reopen. He will describe today not only what it will take to restart the mine, but also his intent to augment America's capability to produce the magnets needed for electrical generators in wind turbines. From what he has told us in preparation for the hearing, he's found it hard to get help at making his vision a reality. If we intend to rebuild America's capability to supply its own needs in rare earth materials, if we intend to foster a home-grown capability to make the devices that provide wind energy, we can't succeed unless he and others like him succeed.

Further, are we investing enough in research looking into ways to recover and recycle these materials and looking for alternatives or synthetic options? Are there efficiencies that could be gained in the use of rare earth materials? For example, if you work with rare earths on the nanoscale level, could you get the same improvements in material performance using micrograms where today you need kilograms? There aren't a lot of places where people are currently working to answer these questions even as the answers could go far in helping America compete in the alternative energy technology industries springing up around the globe.

This is not the first time the Committee has wrestled with rare earth and critical materials issues.

Our Committee established a national policy in minerals and materials three decades ago. That 1980 law required a continuing assessment of mineral and materials markets to alert us to looming problems such as supply disruptions, price spikes and the like.

Four years later we followed up by establishing the Critical Materials Council to assure that someone was minding the store. However, you won't find the Critical Materials Council in the White House organization chart today; it disappeared into the National Science and Technology Council in 1993 and high level attention to rare earths, and other materials, fell away as a priority.

While preparing for this hearing, we have learned that the Office of Science and Technology Policy has recently organized a new interagency committee to respond to our rare earth problems. An obvious question arises: if the Critical

Materials Council had been maintained might we be in a better position to protect our nation's interests in a robust rare earths industry? How can we reverse the result of that history of neglect?

The Subcommittee thanks the witnesses for helping us address these issues and I anticipate an interesting discussion later. I now recognize Dr. Broun, our Ranking Member, for his opening remarks.

In: Rare Earth Minerals: Policies and Issues
Editor: Steven M. Franks

ISBN: 978-1-61122-310-1
© 2011 Nova Science Publishers, Inc.

Chapter 5

STATEMENT OF DR. STEPHEN FREIMAN, PRESIDENT, FREIMAN CONSULTING, BEFORE THE SUBCOMMITTEE ON INVESTIGATIONS AND OVERSIGHT, HEARING ON "RARE EARTH MINERALS AND 21ST CENTURY INDUSTRY"

Stephen Freiman

Good afternoon, Mr. Chairman and members of the Committee. My name is Dr. Stephen Freiman. A few years ago I retired as Deputy Director of the Materials Science and Engineering Laboratory at the National Institute of Standards and Technology to start a small consulting business. I served on the Committee on Critical Mineral Impacts on the U.S. Economy of the National Research Council (NRC). The Research Council is the operating arm of the National Academy of Sciences, National Academy of Engineering, and the Institute of Medicine of the National Academies, chartered by Congress in 1863 to advise the government on matters of science and technology.

Mineral-based materials are ubiquitous—aluminum in jet aircraft; steel in bridges and buildings, and lead in batteries, to name but a few examples. The emergence of new technologies and engineered materials creates the prospect of rapid increases in demand for some minerals previously used in relatively small quantities in a small number of applications—such as lithium in automotive batteries, rare-earth elements in permanent magnets and compact-fluorescent light bulbs, and indium and tellurium in photovoltaic solar cells. At the same time, the supplies of some minerals seemingly are becoming increasingly fragile due to more fragmented supply chains, increased U.S. import dependence, export restrictions by some nations on primary raw materials, and increased industry concentration.

It was in this light that the U.S. Geological Survey (USGS) and the National Mining Association sponsored a National Research Council study to examine the range of issues important in understanding the evolving role of nonfuel minerals in the U.S. economy and the potential impediments to the supplies of these minerals to domestic users. The study was conducted under the purview of the NRC's standing Committee on Earth Resources. The findings of the study are contained in the volume *Minerals, Critical Minerals, and the U.S. Economy* (National Academies Press, 2008).

In my testimony today, I highlight two parts of the report: its analytical framework and empirical findings, and its recommendations. In addition, I provide answers to the questions you posed in your letter of invitation to me.

ANALYTICAL FRAMEWORK

The analytical framework begins by defining critical minerals as those that are both essential in use (difficult to substitute away from) and subject to supply risk. The idea is illustrated in Figure 1, a 'criticality matrix.' The horizontal axis represents the degree of supply risk associated with a particular mineral, which increases from left to right. Supply risk is higher (1) the greater the concentration of production in a small number of mines, companies, or countries, (2) the smaller the existing market (the more vulnerable a market is to being overwhelmed by a rapid increase in demand due to a large new application), (3) the greater the reliance on byproduct production of a mineral (because the supply of a byproduct is determined largely by the economic attractiveness of the associated main product), and (4) the smaller the reliance on post-consumer scrap as a source of supply. Import dependence, by itself, is a poor indicator of supply risk; rather it is import dependence combined with concentrated production and perhaps geopolitical risk (the first of the four factors above) that lead to supply risk. In Figure 1, the hypothetical mineral A is subject to greater supply risk than mineral B.

The vertical axis represents the impact of a supply restriction, which increases from bottom to top. Broadly speaking, the impact of a restriction relates directly to the ease or difficulty of substituting away from the mineral in question. The more difficult substitution is, the greater the impact of a restriction (and vice versa). The impact of a supply restriction can take two possible forms: higher costs for users (and potentially lower profitability), or physical unavailability (and a "no-build" situation for users).[1]

The overall degree of criticality increases as one moves from the lower-left to the upper- right corner of the diagram. The hypothetical mineral A would be relatively more critical than mineral B.

Implementing the framework requires specifying a perspective and time frame. The perspective of a mineral-using company, for example, will likely be different than that of a national government. The degree of criticality in the short to medium term (one or a few years, up to a decade) depends on *existing* technologies and production facilities. Substituting one material for another in a product typically is difficult in the short term due to constraints imposed by existing product designs and production equipment. Short- term supply risks are a function of the nature and location of existing production. In contrast, over the longer term (a decade or more), the degree of criticality depends much more importantly on technological innovation and investments in *new* technology and equipment on both the demand side (material substitution) and the supply side (mineral exploration, mining and mineral processing, and associated technologies).

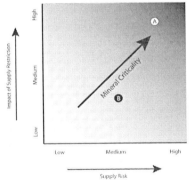

Figure 1. The Criticality Matrix. Source: *Minerals, Critical Minerals, and the U.S. Economy* (National Academies Press, 2008)

Taking the perspective of the U.S. economy overall and in the short to medium term, the committee evaluated eleven minerals or mineral families. It did not assess the criticality of all important nonfuel minerals due to limits on time and resources. Figure 2 summarizes the committee's evaluations. Of the eleven minerals, those deemed most critical—that is, they plot in the upper-right portion of the diagram—are indium, manganese, niobium, platinum-group metals, and rare-earth elements.

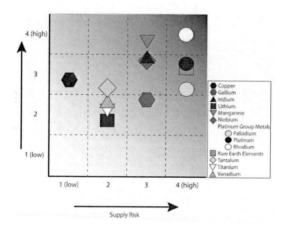

Figure 2. Criticality Evaluations for Selected Minerals or Mineral Families. Source: *Minerals, Critical Minerals, and the U.S. Economy* (National Academies Press, 2008)

A final point: criticality is dynamic. A critical mineral today may become less critical either because substitutes or new sources of supply are developed. Conversely, a less- critical mineral today may become more critical in the future because of a new use or a change in supply risk. Such could be the case with lithium, which the committee did not evaluate as one of the more-critical minerals in its analysis two years ago (Figure 2); if demand for lithium in batteries increases significantly and new sources of supply are in politically risky locations, then lithium could plot in the more-critical region of the figure in the future.

RECOMMENDATIONS

The committee made three recommendations, which I quote below:

1. The federal government should enhance the types of data and information it collects, disseminates, and analyzes on minerals and

mineral products, especially as these data and information relate to minerals and mineral products that are or may become critical.

2. The federal government should continue to carry out the necessary function of collecting, disseminating, and analyzing mineral data and information. The USGS Minerals Information Team, or whatever federal unit might later be assigned these responsibilities, should have greater authority and autonomy than at present. It also should have sufficient resources to carry out its mandate, which would be broader than the Minerals Information Team's current mandate if the committee's recommendations are adopted. It should establish formal mechanisms for communicating with users, government and nongovernmental organizations or institutes, and the private sector on the types and quality of data and information it collects, disseminates, and analyzes. It should be organized to have the flexibility to collect, disseminate, and analyze additional, nonbasic data and information, in consultation with users, as specific minerals and mineral products become relatively more critical over time (and vice versa).

3. Federal agencies, including the National Science Foundation, Department of the Interior (including the USGS), Department of Defense, Department of Energy, and Department of Commerce, should develop and fund activities, including basic science and policy research, to encourage U.S. innovation in the area of critical minerals and materials and to enhance understanding of global mineral availability and use.

QUESTIONS FROM THE SUBCOMMITTEE ON INVESTIGATIONS AND OVERSIGHT

What are the Major Gaps in Current Federal policy for Minerals and Materials?

The committee report does not address this broad question. It does identify gaps in minerals information and recommends enhanced collection, dissemination and analysis of those parts of the mineral life cycle that are under-represented at present including: reserves and subeconomic resources, byproduct and coproduct primary production, stocks and flows of materials available for recycling, in-use stocks, material flows, and materials embodied in internationally traded goods. The committee report recommends periodic analysis of mineral criticality over a range of minerals.

Which Aspects of Research and Development in Minerals and Materials Require Enhanced Federal Support, and What Form Should This Support Take?

See Recommendation 3 above. As part of its detailed discussion of this recommendation, the committee report also recommends funding scientific, technical, and social-scientific research on the entire mineral life cycle. It recommends cooperative programs involving academic organizations, industry, and government to enhance education and applied research.

How Should the Federal Government Improve the Collection of Information on Minerals and Materials Markets?

See Recommendation 2 above. As part of its more detailed discussion of this recommendation, the committee report suggests that the Federal government consider the Energy Information Administration, which has status as a principal statistical agency, as a potential model for minerals information, dissemination, and analysis. Whatever agency or unit is responsible for minerals information, it needs greater autonomy and authority than at present.

Facing Dynamic Changes in Supply and Demand for Particular Minerals and Materials in a Global Economy, What Are the Most Useful Contributions the Federal Government Can Employ to Assist Industry?

My personal opinion is that federal minerals and materials policy should focus on: (1) encouraging undistorted international trade, (2) ensuring that policies and procedures for domestic mineral development appropriately integrate commercial, environmental, and social considerations, (3) facilitating provision of information on which private and public decisions are made, and (4) facilitating research and development, including on recycling of specialty materials used in small quantities in emerging uses.

Thank you for the opportunity to testify today. I would be happy to address any questions the subcommittee may have.

DR. STEPHEN FREIMAN

Dr. Freiman graduated from the Georgia Institute of Technology with a B. ChE. and a M. S. in Metallurgy. After receiving a Ph.D. in Materials Science and Engineering from the University of Florida in 1968, Dr. Freiman worked at the IIT Research Institute and the Naval Research Laboratory. He joined NIST (then NBS) in 1978. From 1992 to 2002 Dr. Freiman served as Chief of the Ceramics Division at NIST. Prior to his leaving NIST in 2006 to start a consulting business (Freiman Consulting Inc.), Dr. Stephen Freiman served for four years as Deputy Director of the Materials Science and Engineering.

Dr. Freiman has published over 200 scientific papers focusing primarily on the mechanical properties of brittle materials. He was the first Chairman of the ASTM Subcommittee addressing brittle fracture and a past Chair of the Steering Committee of the Versailles Project for Advanced Materials and Standards. Dr. Freiman served as Treasurer, and President of the American Ceramic Society, and is a Fellow and Distinguished Life Member of the Society.

End Notes

[1] When considering security of petroleum supplies, rather than minerals, the primary concern is costs and resulting impacts on the macroeconomy (the level of economic output). The mineral and mineral-using sectors, in contrast, are much smaller, and thus we are not concerned about macroeconomic effects of restricted mineral supplies. Rather the concern is both about higher input costs for mineral users and, in some cases, physical unavailability of an important input.

In: Rare Earth Minerals: Policies and Issues   ISBN: 978-1-61122-310-1
Editor: Steven M. Franks   © 2011 Nova Science Publishers, Inc.

Chapter 6

STATEMENT OF KARL A. GSCHNEIDNER, JR., AMES LABORATORY, U.S. DEPARTMENT OF ENERGY AND DEPT. OF MATERIALS SCIENCE AND ENGINEERING, IOWA STATE UNIVERSITY, BEFORE THE SUBCOMMITTEE ON INVESTIGATIONS AND OVERSIGHT, HEARING ON "RARE EARTH MINERALS AND 21ST CENTURY INDUSTRY"

*Karl A. Gschneidner**

QUESTION 1: RARE EARTH SCIENCE AND TECHNOLOGY AT THE AMES LABORATORY

1. Introduction

The Ames Laboratory (AL) is the smallest of DOE's (Department of Energy) seventeen national laboratories. It is a single-program laboratory with

* E-mail: cagey@ameslab.gov, Phone: 515-294-7931

about 78% of DOE's funding from the Office of Basic Energy Sciences (BE S). Additional non-DOE income of about $7M is derived from contracts, grants, Cooperative Research and Development Agreements (CRADAs) and Work for Others (WFO) arrangements.

The AL is fully integrated with Iowa State University (ISU) and its buildings are right on the campus and several are directly connected with ISU buildings. All AL personnel are ISU employees, and many of the lead scientists (23) have joint appointments with various academic departments. There are about 140 scientists and engineers, 260 graduate and undergraduate students, another 240 visiting scientists, facility users, and associates, and 180 support personnel, for a total of about 440 employees (or about 300 full-time equivalent employees) and 410 associates (non-payroll). Of the scientific staff about 20% are directly involved in rare earth research and development activities, including materials science, condensed matter physics, and materials chemistry.

2. A Brief History – How Did Rare Earths Get to Ames?

Many persons may wonder how the Mecca of rare earths[a] ever ended up on the picturesque ISU campus – an oasis amongst the corn and soy bean fields of central Iowa. The story begins in late 1930s when Frank H. Spedding was searching for a permanent academic position after receiving his Ph.D. from the University of California in 1929 under G.N. Lewis. His Ph.D. thesis dealt with the physical properties of rare earth materials at low temperature. He spent a number of years in temporary jobs, including two years in Europe, where he worked with several Nobel prize winners, including Niel Bohr. In 1937, while he was occupied at Cornell University working with a future Nobel prize winner, Hans Bethe, he was offered a permanent position at ISU[b] as an associate professor, and he remained at ISU until he retired in 1972. Frank Spedding was primarily a spectroscopist, but he had to separate and purify his rare earth samples from the other rare earth elements in order to carry out his optical measurements, which are very sensitive to the presence of other rare earth impurities. Based on his experience with rare earth (or $4f$) chemistry he was asked by A.H. Compton of the University of Chicago to work with actinide (or $5f$) materials to assist the team of physicists to build the first nuclear reactor under the stands of Stagg Field at the University of Chicago. Spedding was appointed the Director of Chemistry in the Chicago Project. One of the main goals of the Project was to purify uranium from its ores and then to convert the oxide to the metal – six tons were needed, which seemed like mission impossible since only a few grams of uranium metal had been produced before World War II. By late summer in 1942 the delivery of the uranium was behind schedule and Frank Spedding offered to use a different approach to make the metal via a metallothermic reduction of uranium tetrafluoride, UF_4. It was successful, and as an added bonus it was purer than the metal produced by other organizations. By November, the ISU team sent two tons of uranium cylinders (2" diameter x 2" long) to Chicago. The addition of these two tons to the four tons delivered previously allowed the reactor to go critical on December 2, 1942. This stellar contribution to the Manhattan Project was recognized when ISU and Spedding's team were awarded the "Army and Navy E for Effort with four stars" pennant on October 12, 1945; the only university (college) in the country to receive this prestigious honor. Since the Ames ingots were purer and less expensive to produce, the Manhattan District asked three companies to take over the Ames fluoride process and make large amounts of uranium for the Oak Ridge and Hanford reactors. But since it took time to get the facilities up and running the ISU group was asked to continue to produce uranium metal. More than two million pounds were produced by the end of World War II, and by December 1945 industry took over. The Ames process, albeit somewhat modified, is still in use today [1].

Soon thereafter, a new requirement arose in the war effort. Thorium metal was needed for another type of reactor, and Spedding and his co-workers were asked to develop a process for making thorium metal. Again they were successful and the Ames thorium process was turned over to industry. In the meanwhile, 600,000 pounds of thorium were produced [1].

During the late war years, research on developing methods of separating the rare earth elements was begun at several Manhattan District laboratories because some of the rare earth elements were among fission products which would absorb neutrons because of their high neutron capture cross-section and eventually would cause the reactor to shut down. Also the pure rare earth elements were needed to study their chemical and metallurgical behaviors to help actinide chemists and physicists understand the trans-uranium elements because of the expected similarities in the properties of the two series of elements. It was during these years that Spedding and co-workers developed the ion exchange method for separating rare earths, which was in commercial use for many years after World War II, until displaced by liquid- liquid extraction procedures. The ion exchange process is still in use today to obtain the very highest purity rare earth elements (99.9999% pure), which are primarily used in optical applications such as lasers and optical signal multipliers [1].

On May, 17, 1947 the Ames Laboratory became one of the Atomic Energy Commission (AEC) laboratories to promote the peaceful uses of atomic energy and to do research to increase our understanding and knowledge of the chemistry, physics, and nuclear behavior of the lesser known and uncommon elements, including the rare earths [1].

Additional information about the Ames Laboratory and Frank H. Spedding during World War II and the early post-World War II years can be found in several articles authored by I.E. Goldman [2,3,4] and papers by J.D. Corbett [5] and by S.R. Karsjen [6].

3. The Golden Age of Rare Earth Research

The time span of the 1950s through the 1970s was the golden age of rare earth research at the Ames Laboratory. At that time there was a very wide spectrum of research being carried out ranging from separation chemistry to analytical chemistry to process and physical metallurgy to solid state experimental physics to theoretical first principle calculations.

3.1. Separation Chemistry

The discovery of using ion exchange chromatography to separate and purify the rare earths was further refined and improved in the 1950s through the 1960s. A large pilot plant was set up to supply researchers at the Ames Laboratory and other organizations, including many of the AECs national laboratories, with high purity individual rare earths, to carry out fundamental and applied research on various chemical compounds and the pure metals (see § §3.3-3.6). In addition these ion exchange columns were used to separate the other rare earths from yttrium (also a rare earth element) which was to be used in the nuclear aircraft (see § 3.3).

3.2. Analytical Chemistry

In order to verify the chemical purity of the separated products new and more sensitive chemical and physico-chemical analytical methods were developed to detect impurities of both rare earth and other non-rare earth elements at the part per million level. This included wet chemistry, atomic emission and atomic absorption spectroscopy, laser ion mass spectrometry, vacuum and inert gas fusion, and combustion analysis. This research also led to the development of inductively coupled plasma (ICP)-atomic emission (AE) in the late 1970s, and ICP-mass spectrometry (MS), which occurred in the early 1980s. The ICP-MS technology was turned over to industry and today is still one of the most versatile and utilized analytical techniques, see § 5.

3.3. Process Metallurgy

This was one of the strengths of the Ames Laboratory in this time period. The pure rare earth elements, after separating them on the ion exchange columns, were converted to their respective rare earth oxides. The oxide was converted to the fluoride which was then reduced to the pure metal by calcium metal. These two processes were the critical steps for preparing high purity metals with low concentration of interstitial impurities, especially oxygen, carbon, nitrogen, and hydrogen. The reduced metals were further purified by a vacuum casting step and for the more volatile rare earth metals further purification was carried out by distillation or sublimation. Generally, kilogram (2.2 pounds) quantities were prepared at the Ames Laboratory. Industry adopted the Ames process with some minor modifications to prepare commercial grade rare earth metals, and it is still in use today. This method is still being carried out at the AL under the auspices of the Materials Preparation Center (MPC), see § 6.1.

In the late 1950s the AEC asked the Ames Laboratory to prepare pure yttrium metal for the proposed nuclear aircraft. A nuclear reactor would be used to heat gases to propel an airplane much like a jet engine. This aircraft would carry atomic weapons for months at a time without landing. The yttrium was to be hydride to form YH_2 which is used to absorb the neutrons produced by the fission of uranium protecting the crew from radiation. As part of this project the AL produced 65,000 pounds of YF_3 and 30,000 pounds of yttrium metal. The yttrium metal was cast into 85 pound, 6 inch diameter ingots to ship to the General Electric Co. facilities in Cincinnati, Ohio.

In the mid-1950s, Spedding and A.H. Daane and their colleagues developed a new technique for preparing high purity metals of the four highly volatility rare earths – samarium, europium, thulium, and ytterbium – by heating the respective oxides with lanthanum metal and collecting the metal vapors on a condenser. This process has also been turned over to industry and is used by AL's Materials Preparation Center today (§ 6.1).

3.4. Physical Metallurgy

Research in the physical metallurgy area encompassed: determining melting points, crystal structures, vapor pressures, low temperature heat capacities, elastic constants, and magnetic properties of the pure metals and various intermetallic compounds; and phase diagram and thermodynamic properties studies; crystal chemistry analyses; and alloying theory of rare earth-based materials. This was another strong focus area in the AL in the 1950s-1970s era which is still active today but at reduced level.

Closely related, but not strictly physical metallurgy, was research on the mechanical behavior (tensile and yield strengths, ductility, and hardness) of the metals and some of their alloys, and also oxidation and corrosion studies, especially yttrium in conjunction with the nuclear aircraft project. From the information gained from the mechanical property measurements, processes were developed to fabricate the rare earth metals and their alloys at room and elevated temperature into a variety of shapes and forms, e.g. rolled sheets.

3.5. Materials and Solid State Chemistry, and Ceramics

The chemical activity, in addition to separation and analytical chemistry §3.1 and §3.2) was another area in which the AL was considered to be world class, even though the manpower levels were smaller than above noted areas of analytical chemistry, process and physical metallurgy. Research was focused on the sub-stoichiometric rare earth halides, and interstitial impurity stabilized compounds including the halides. X-ray crystallography was an important tool in this focus area to characterize these compounds. John D. Corbett was the lead scientist and is still active today.

Investigations of ceramic materials were important in many of the studies and advances in process and physical metallurgy, not only for their refractory properties, but also because of the need to contain the molten metals without contamination. Rare earth oxides and sulfides, because of their intrinsic stability, were candidate materials to contain the molten rare earths, uranium, thorium and other non-rare earth metals.

3.6. Condensed Matter Physics

The AL was very strong in this area from the very beginning and still is today. Research under the leadership of Sam Legvold was concentrated on the magnetic behavior of the metals and the intra-rare-earth alloys. This work was strongly coupled with neutron scattering studies at both the Ames Laboratory and Oak Ridge National Laboratory, and also with the theorists. The theoretical efforts included first principle calculations (Bruce Harmon) and phenomenological approaches (Sam Liu). Superconductivity was also another active topic of research, but most of the effort was concentrated on non-rare-earth compounds.

3.7. Interdisciplinary Research

Two of the main strengths of the Ames Laboratory are magnetism and X-ray crystallography of rare earth and related materials. In part this is due to cooperative research efforts that cut across the disciplines of physics, materials science, and chemistry. Frank Spedding was one of the leaders in this approach to scientific research, which was rare in the 1950s.

4. Interactions with Industry

As mentioned above in §3 much of the research and development efforts were turned over to industry – the uranium and thorium metal production, the ion exchange separation processes, and the analytical techniques (especially ICP-MS). In addition, K.A. Gschneidner, Jr. established the Rare-earth Information Center (RIC) in 1966 with the initial support of the forerunner of BES, and later by industry (starting in 1968), which totaled about 100 companies world-wide in 1996. RIC's mission was to collect, store, evaluate and disseminate information about new scientific discoveries, industrial developments, new commercial products, conferences, books and other literature, honors received by rare earthers, and to answer information inquiries. RIC published two newsletters – a quarterly (available free) and a monthly (available to supporters of RIC), and occasional reports. In 1996 the directorship was turned over to R.W. (Bill) McCallum. But RIC ceased operation in August 2002 – when industry support dwindled significantly as China forced many companies out of the rare earth markets with extreme price reductions and, simultaneously, a down-turn in the economy dried up state and federal support.

Because of the expertise of individual AL scientists and/or some unique AL analytical or processing capability, many organizations, including industrial companies, asked the AL to perform applied research as Work for Others projects or CRADAs. Many of these non-DOE projects include the rare earths. One of these cases is discussed in more detail in § 6.5. In addition to these individual interactions, the AL established the Materials Preparation Center

(MPC) in 1981 to provide unique metals, alloys and compounds to worldwide scientific and industrial communities; and to perform unusual processes for fabricating materials which could not be done elsewhere, see § 6.1. The functions carried out by the MPC over the nearly 30 years of its existence are an outstanding example of AL-industry interactions.

Over the years various industrial organizations have sent their staff scientists and engineers to work at Ames Laboratory getting firsthand experience on a particular technology. These arrangements may be part of CRADAs or Work for Others projects.

In 2009 ISU became a research member of the Rare Earth Industry and Technology Association (REITA) to implement rare earth technology and promote commercialization of the rare earths for military and civilian applications.

5. Technology Transfer and Patents

The Ames Laboratory (AL) has been awarded 300 patents, of which about 45 are concerned with rare earth materials. Ten patents deal with the rare earth-base permanent magnets and four with magnetic refrigeration materials. Before the passage of the Stevenson-Wydler Technology Innovation Act of 1980 (P.L. 96-480) and the Dayh-Dole Act of 1980 (P.L. 96-5 17) all patent rights were turned over to the U.S. Government. After the Acts became law, the AL (a GOCO - government owned, contractor operated laboratory) began to license the various technologies developed at the AL via the Iowa State University Research Foundation (ISURF).

The combination of inductive coupled plasma with atomic emission spectroscopy in 1975 and later with mass spectrometry in 1984 was a quantum jump in increasing the sensitivity for detecting and determining trace elements in various materials. The two analytical methods were developed at the AL to improve the speed and lower the limits of detecting various rare earth impurity elements in a given rare earth matrix. This technique was soon applied to other impurities in a variety of non-rare-earth materials, e.g. detection of poisons such as mercury and arsenic in drinking water. The ICP-MS and ICP-AE technologies were turned over to industry and are now a standard analytical tool in over 17,000 analytical laboratories worldwide. It is a rapid and accurate method for 80 elements, and in some cases allows the detection of an impurity down to the parts per trillion level. Today there are at least six companies that manufacture ICP-MS instruments.

In addition to the various technology transfers noted in the previous paragraph and in sections 2, 3.1-3.3, one of the more recent success stories is concerned with Terfenol. Terfenol is a magnetic iron-rare-earth (containing dysprosium-terbium) intermetallic compound which has excellent magnetostrictive properties. When a magnetic field is turned on Terfenol will expand and when the magnetic field is removed it relaxes to its original shape and size. There are many applications for this material including sonar devices for detecting submarines, oil well logging, vibration dampers, audio speakers, etc. The magnetostrictive properties were discovered in the early 1970s at the Naval Ordinance Laboratory in Maryland. Shortly thereafter the Navy contracted the AL to grow single crystals and Terfenol samples with preferred orientations. The AL was successful and designed a procedure for making the orientated material to maximize the amplitude of the magnetostrictive effect. Patents were issued and in the late 1980s ISURF licensed the processing technology to Etrema, a subsidiary of Edge Technologies, Inc. in Ames, Iowa. Today Etrema is a multimillion dollar business.

6. Where We Are Today 1980-2010

With the Ames Laboratory's successes, some of the golden-age research was no longer deemed to be basic research and funding dried up. In addition key personnel started to retire. As a result of these two events a number of AL capabilities were phased out completely. These include: analytical chemistry, separation chemistry, process metallurgy, and ceramics. The excellent analytical capabilities were slowly reduced and completely lost by the 2000s, except for inert gas fusion and combustion analysis. The rare earth research activities in physical metallurgy and condensed matter physics areas have also suffered some downsizing to about half the level of what it was in the pre-1980 era, but what is left is still first class state-of-the-art basic research.

In the following sections important activities that are still ongoing are described. Other research that had been completed in the 1980s and may play in important role in the future activities of a new national rare earth research center, is also noted.

6.1. Materials Preparation Center

As an outgrowth of the Ames Laboratory's interactions with industry, other DOE laboratories, universities, other research organizations, the Materials Preparation Center (MPC) was established in 1981 to provide high purity metals (including the rare earths, uranium, thorium, vanadium, chromium); and intermetallics, refractory, and inorganic compounds, and specialty alloys; none of which are available commercially in the required purity or form/shape needed by the requestor on a cost recovery basis. The MPC is a BES specialized research center with unique capabilities in the preparation, purification, processing, and fabrication of well-characterized materials for research and development. The Center is focused on establishing and maintaining materials synthesis and processing capabilities crucial for the discovery and development of a wide variety of use-inspired, energy-relevant materials in both single crystalline and polycrystalline forms, spanning a range of sizes with well-controlled microstructures. There are four functional sections within the MPC: (1) high purity rare earth metals and alloys; (2) general alloy preparation; (3) single crystal synthesis; and (4) metallic powder atomization. Each area is provided scientific and technical guidance by a Principal Investigator (PI) whose individual expertise is aligned with the function of each section. The original director was F. (Rick) A. Schmidt who retired in 1993 and turned over the directorship to Larry L. Jones.

In 2008 the MPC filled 183 external materials requests from 111 different scientists at 88 academic, national and industrial laboratories worldwide. Internally the Center provided materials, and services for 53 different research projects that totaled 1092 individual requests.

6.2. Nd$_2$Fe$_{14}$B Permanent Magnets

The announcement of the simultaneous discovery of the high strength permanent magnet materials based on Nd$_2$Fe$_{14}$B by scientists at General Motors in the USA and at Sumitomo Special Metals. Co., Ltd. in Japan in November of 1983 set off a flurry of activities everywhere. The lead scientist at General Motors was John Croat (an ISU graduate), who was Frank Spedding's last graduate student. DOE/BES funding for research on these materials at the AL started in 1986 and lasted through 1998. U.S. Department of Commerce (DOC) funding for gas atomization processing work on Nd$_2$Fe$_{14}$B alloys, through the ISU Center for Applied Research and Technology, was received from 1988 through 1993. Funding was renewed at the AL in 2001 under the auspices of DOE/EERE's Vehicle Technology (formerly FreedomCar) program.

Notable achievements in the BES funded project included: (1) demonstrating that the Nd2Fe14B compound can be prepared by a thermite reduction process that is competitive with other methods of the permanent magnet material; (2) developing methods for controlling the solidification microstructure of melt spun $Nd_2Fe_{14}B$ which leads to large energy products (the larger the energy product the better the permanent magnet properties); (3) proposing a model for the rapid solidification of a peritectic compound to explain the solidification microstructure of melt spun $Nd_2Fe_{14}B$; and (4) developing a model for hysteresis in exchange coupled nanostructure magnets. In 1996 the AL team headed by R.W. (Bill) McCallum[c] received the DOE Materials Science Award for "Significant Implications for DOE Related Technologies, Metallurgy and Ceramics" (items 2 and 3 above). A year later this same team won an R&D-100 Award for Nanocrystalline Composite Coercive Magnet Powder (see § 7.1). In the DOC funded project, an alternative rapid solidification process, gas atomization, was developed for making fine spherical $Nd_2Fe_{14}B$ powders, for which the AL (Iver Anderson and Barbara Lograsso) received an R&D-100 award in 1991 (see § 7.1). They also received the Federal Laboratory Consortium Award for Excellence in Technology Transfer for gas atomization processing of $Nd_2Fe_{14}B$ to enable improved molding of bonded magnets. The AL thermite reduction process (item 1), which was developed by F. (Rick) A. Schmidt, J.T. Wheelock and Dave T. Peterson under MPC research, was selected for one of the 1990 IR-100 (changed to "R&D 100" in 1991) Awards for new innovative research for potential commercialization.

The Vehicle Technology research funded by EERE is on-going and includes design of improved $Nd_2Fe_{14}B$ permanent magnets which can operate at high temperature, enabling more powerful and more efficient motors. This project also is developing further the high temperature RE magnet alloys for powder processing, intended for injection molded bonded magnets for mass production of hybrid and electrical vehicles. Based on initial success with both aspects of the RE magnet project, in 2009 EERE expanded their support into the high risk task of identifying non-rare-earth magnet alloys with sufficient strength for vehicle traction motors.

6.3. $Nd_2Fe_{14}B$ Scrap Recovery

As manufacturers began to make the $Nd_2Fe_{14}B$ material, it soon became apparent there was a great deal of waste magnet material being generated because grinding, melting and polishing the magnets into a final form/shape. Much of the magnet material is mixed with oils and other liquids used in these operations – this material is known as "swarf". The team of scientists at AL headed by F. (Rick) A. Schmidt developed two different processes to recover the neodymium metal: a liquid metal extraction process to treat the solid materials; and an aqueous method for treating the swarf. Both processes were patented, but the patents have since expired.

6.4. High Temperature Ceramic Oxide Superconductors

In the mid-1980s another major discovery occurred and had an enormous impact on the rare earths as well as science and technology in general – the discovery of the oxide superconductor with transition temperatures greater than that of liquid nitrogen 77 K (-195°C). One of the key superconductors was $YBa_2Cu_3O_7$, also known as "1:2:3". It is utilized today in electrical transmission lines, electrical leads in low temperature high magnetic field apparati and other superconducting applications. The AL had a strong tradition in superconducting research well before this discovery, and when they learned of it the condensed matter physicists and materials scientists immediately began research on these ceramic oxide superconductors. A National Superconducting Basic Information Center was established at AL in 1987 with financial support from DOE's BES. It was headed by John R. Clem, a theorist who continues to consult with American Superconductor. The experimentalists worked diligently on various aspects of the 1:2:3 and other oxide superconductors to understand the processes by which they are formed and to prepare high purity well characterized materials for physical property studies, which would assist the theorists to understand the fundamental nature of these superconductors. This work laid the ground work for the development of a method of fabricating the rare earth 1:2:3 materials into filaments and flexible wires. Most of the research on these oxide superconductors at AL has stopped and most of the know-how has been turned over to industry. However, AL scientists are still at the forefront of the field studying the new high temperature superconductors: the rare-earth-arsenic-iron-oxide-fluoride and the MgB_2 materials.

6.5. Magnetic Cooling

Magnetic cooling is new, advanced, highly technical, energy efficient, green technology for cooling and climate control of buildings (large and homes), refrigerating and freezing food (supermarket chillers, food processing plants, home refrigerator/freezers). The AL team headed by K.A. Gschneidner, Jr. and V.K. Pecharsky has been involved with magnetic cooling since 1990, when Astronautics Corporation of America (ACA) asked Gschneidner to develop a new magnetic refrigerant material to replace the expensive GdPd refrigerant they were using for hydrogen gas liquefaction (a DOE sponsored research effort). The AL team was successful and showed that a $(Dy_{0.5}Er_{0.5})Al_2$ alloy was about 1000 times cheaper and 20% more efficient than GdPd. A patent was issued for this new magnetic refrigerant material. This work was recognized as the best research paper presented at the 1993 Cyrogenic Engineering Conference. A few years later AL teamed up with ACA and designed, constructed and tested a near room temperature magnetic refrigerator. In 1997 they demonstrated that near room temperature magnetic refrigeration is competitive with conventional gas compression cooling technology and is about 10% more efficient, and is a much greener technology because it does not employ ozone depleting, or greenhouse, or hazardous gases [7]. This work was funded by BES's Advanced Energy Project program. Additional research on magnetocaloric materials was supported by BES after the Advanced Energy Project ended in 1998. But in 2005 BES funding for this research was terminated because they thought it was no longer basic research, i.e. it was too applied. Since then some work has continued on magnetocaloric materials under a work for others subcontract with ACA who has a Navy contract to build shipboard cooling machines, and a few SBIRs which are being funded by EERE.

This research on magnetic cooling is a good example of AL's response to a problem encountered by industry which was successfully solved, and then later, this work led to a whole new cooperative AL-industry project on near room temperature magnetic refrigeration.

6.6. Neutron Scattering

Neutron scattering is a powerful tool in determining magnetic structures of magnetic materials and it compliments magnetic property measurements made by standard magnetometers. The rare earth research at AL has benefited from interactions with the neutron scatterers. In the early 1950s Frank Spedding and Sam Legvold of the AL had a close relationship with the neutron scattering group headed by Wally Koeller at Oak Ridge National Laboratory neutron scattering facility and furnished single crystals of the rare earth metals. Recognition and demand for neutron scattering resulted in a 5MW reactor being constructed locally for Ames Laboratory. Scientists used this reactor for extensive measurements of the electronic interactions in rare earth and other magnetic materials. Because of a large jump in the cost of operating and fueling this reactor, it was shut down in 1978. The relationship with the neutron scattering effort at Oak Ridge was enhanced and continued for many years up to about 1980, shortly before the death of the three scientists in 1983-84. To this day a dedicated neutron scattering facility, run by AL scientists, operates at the Oak Ridge High Flux Isotope Reactor (HFIR). It is still of great benefit to AL scientists studying rare earth materials.

6.7. X-ray Magnetic Scattering

X-ray magnetic scattering is a fairly new tool, which was developed in the early 1990s, to study magnetic structures. It is fortunate that this new tool became available because a few of the rare earth elements, especially gadolinium, readily absorb neutrons and neutron scattering measurements are very difficult if impossible to make. Thus, X-ray magnetic scattering has been especially useful in determining the magnetic structures of gadolinium compounds.

In more recent years scientists have improved the X-ray magnetic scattering technique, which is called X-ray magnetic circular dischroism (XMCD). The AL scientists have been on the forefront by applying the latest experiments and theoretical tools to help elucidate complex electronic interactions underlying bulk magnetic properties. The AL team, led by Alan Goldman (experiment) and Bruce Harmon (theory), has been pioneers in the development and application of XMCD on rare earth materials. This tool gives valuable and direct information about the itinerant electrons responsible for coupling the individual localized magnetic moments of each rare earth atom in a solid. The stronger the microscopic coupling the stronger the bulk magnet, and the more useful it can be in applications. Such experiments and powerful computers are essential for helping AL scientists in their latest "materials discovery" initiative to accelerate the discovery of new magnetic materials for industry.

6.8. Emerging Technologies

One of the new and exciting, ongoing developments at Ames Laboratory is a revolutionary method of preparing rare earth-based master alloys for energy and other applications. In addition to lowering costs of the starting material, the processing technique also reduces energy consumption by 40 to 50% and is a very green technology. The work on preparing $Nd_2Fe_{14}B$ magnet material began about a year ago with financial support from AL patent royalties, and it has been reduced to practice – we have prepared a state-of-the-art permanent magnet on February 5, 2010, see attached figure. It is a one step process going from the neodymium oxide to the neodymium master alloy, and since the end-products are completely utilized, there are no waste materials to dispose of. The conventional process also starts with the neodymium oxide but takes two steps to obtain the neodymium metal, and there are waste products associated with both steps which need to be disposed of in an environmentally friendly manner. The step to prepare the $Nd_2Fe_{14}B$ magnet material is essentially the same in either case. This processing technique was invented by F. (Rick) A. Schmidt and K.A. Gschneidner, Jr. A provisional patent has been filed.

A modification of this process should enable us to prepare a lanthanum master alloy to prepare lanthanum nickel metal hydride batteries, which are used in hybrid and electrical vehicles. Likewise, we believe this process can be used to make magnetic rare earth refrigerant alloys (see § 6.5).

7. Kudos

The Ames Laboratory scientific achievements and their science/engineering leaders have been recognized by several organizations including DOE.

7.1. R&D-1 00 Awards (former IR-1 00 Awards)

Industrial Research magazine annually identifies the nation's top 100 technological innovations, called the IR-100 Awards before 1991 and now are called the R&D-100 Awards. These awards are also known as the "Oscars of Science". Over the past 25 years AL has received 17 R&D-100 Awards. Of these three are involved with rare earths, in particular the $Nd_2Fe_{14}B$ permanent magnet materials. These are listed below.

1990: "Thermite Reduction Process to Make Rare-earth Iron Alloys"
        F. (Rick) A. Schmidt, John T. Wheelock and Dave T. Peterson
1991: "HPGA (High Pressure Gas Atomization)"
        Iver Anderson and Barbara Lograsso
1997: "Nanocrystalline Composite Coercive Magnet Powder"
        R.W. (Bill) McCallum, Kevin Dennis, Matt Kramer, and Dan Branagan

Left: Our KAA-1-34 composition. 60/40 by vol. Nd2Fe14B/PPS (poly(phenylene
    sulfide). Hot pressed at 300°C and magnetized with a 2T electromagnet. The second
    bonded permanent magnet prepared.

Center: Practice magnet of similar composition. The surface is boron nitride coating from
    the die used to compact the $_{Nd2Fe14B}$ particles in the polymer.

Right: First Bond permanent magnet. 30/70 by vol. Nd2Fe14B/diallyl phthalate sample
    mounting material. Hot pressed and sealed with thin layer of epoxy.

Figure Caption: The first and second Nd2 $_{Fe14B}$ bonded permanent magnet prepared using
the new process for making the neodymium master alloy

7.2. National Academies Members

Six Ames Laboratory scientists have been named to the National Academy
of Sciences and the National Academy of Engineering. Frank H. Spedding was
elected in 1952 and John D. Corbett in 1992 to the National Academy of
Sciences. The four National Academy of Engineering members are: Donald O.
Thompson – 1991, Dan Schechtman – 2000, R. Bruce Thompson – 2003, and
Karl A. Gschneidner, Jr. – 2007. Of the six three (Spedding, Corbett and
Gschneidner) were heavily involved in the rare earth science and technology of
rare earths during their careers. Corbett and Gschneidner are still actively

engaged in research and development activities. Spedding died in 1984 but was still active until shortly before his passing.

7.3. Department of Energy Awards

Scientists at AL have won several DOE, (mostly from BES) awards for their scientific achievements. These are listed below.

1982	K.A. Gschneidner, Jr. and K. Ikeda for quenching of spin fluctuations
1991	I.E. Anderson and B.K. Lograsso received the Federal Laboratory Consortium Award for Excellence in Technology Transfer for high pressure gas atomization of rare earth permanent magnet alloys
1994	B.J. Beaudry for thermoelectric materials characterization from DOE's Radioisotope Power Systems Division
1995	J.D. Corbett for sustained outstanding research in materials chemistry
1995	A.I. Goldman, M.J. Kramer, T.A. Lograsso, and R.W. McCallum for sustained outstanding research in solid state physics
1996	D. Branagan, K.W. Dennis, M.J. Kramer, R.W. McCallum for studies on the solidification of rare earth permanent magnets
1997	K.A. Gschneidner, Jr. and V.K. Pecharsky for contributions to the advancement of magnetic refrigeration
2001	K.A. Gschneidner, Jr. and V.K. Pecharsky received the "Energy 100 Award" for research on magnetic refrigeration as one of the 100 discoveries between 1997 and 2000 that resulted in improvements for American consumers.

QUESTION 2: BASIC RESEARCH PROGRAM

In the 1990s the Chinese flooded the marketplace with low priced raw rare earth products (mixed and separated rare earth oxides) and as a result, not only did the primary rare earth producers in the United States and the rest of the world shut down, but technical personnel with expertise in rare earth mining, refining, extraction, etc. found employment in other industries. Soon thereafter the Chinese began manufacturing higher value rare earth products, including rare earth permanent magnet materials, and in time, all of the $Nd_2Fe_{14}B$ magnet manufacturers in the United States also went out of business. This also resulted in a brain drain of scientists and engineers in this field, and also in all high-tech areas involving other rare earth products, such as high energy product permanent magnet materials, metallic hydrogen storage and rechargeable battery materials. Some of these experts have moved on to other industries, others have retired, and others have died, basically leaving behind an intellectual vacuum.

In the late 2000s the Chinese game plan changed, and they have started to exercise export controls on a variety of rare earth products, and signaled that they intend to consume all the rare earths mined in China internally in the next three to five years. This change will allow the rare earth producers and manufacturers to supply the needed products, but this presents several problems which have been cited by others at this House Committee hearing. One of these is the shortage of trained scientists, engineers, and technicians. Another need is innovations in the high tech areas which are critical to our country's future energy needs. A research center which alleviates both of these problems is the best way to work our way through the rare earth crisis facing the USA. An educational institution which has a long and strong tradition in carrying out research on all aspects of rare earth materials – from mining and purification to basic discovery and applications – over a number of disciplines (i.e. chemistry, materials, physics, and engineering) with a strong educational component (undergraduate, graduate and post doctoral students) would be the ideal solution. A National Research Center on Rare Earths and Energy should be established at such an institution initially with federal and, possibly, state support, and as the US rare earth industry matures in five to ten years, supplemented by industrial financial support. The center would employ about 30 full time employees – group leaders; associate and assistant scientists and engineers; post docs, graduate and undergraduate students; and technicians plus support staff. This research center will be a national resource for the rare earth science, technology and applications, and therefore, it would also provide broad support of research activities at other institutions (universities, national laboratories, non-profit research centers, and industry) who would supply intellectual expertise via subcontracts to complement the activities at the center.

The major emphasis of the center would be goal oriented basic research, but proprietary research directly paid by the organizations that request it would also be part of the center's mandate. The center would have an advisory board to oversee, guide and refocus as needed the research being conducted. The advisory board would be made up of representatives from the university, government, industry and the general public.

I would like to suggest to this House subcommittee that they consider a second national center, the National Research Center for Magnetic Cooling. Cooling below room temperature accounts for 15% of the total energy consumed in the USA. As noted in my response to the first question, magnetic refrigeration is a new advanced, highly technical, energy efficient green technology for cooling and climate control of buildings, ships, aircraft, and refrigerating and freezing (§ 6.5). We have shown that magnetic cooling is a refrigeration technology competitive with conventional gas compression cooling. Magnetic cooling is 10 to 20% more efficient, and it is a very green technology because it eliminates hazardous and greenhouse gases, and reduces energy consumption. If we were able to switch all of the cooling processes to magnetic refrigeration at once we would reduce the nation's energy consumption by 5%. But there are a lot of hurdles that need to be overcome and the USA needs to put together a strong, cohesive effort to retain our disappearing leadership in this technology, by assembling a National Research Center for Magnetic Cooling. Europe and China are moving rapidly in this area, and Denmark has assembled a magnetic refrigeration national research center at Risø – so far the only one in the world. The US Center should be structured similar to what has been proposed in the above paragraphs for the National Research Center on Rare Earth and Energy. The question is, are we going to give up our lead position and be a second rate country, or will we be leading the rest of the world? I hope and pray that the answer is, we are going to show the world that we are number one.

QUESTION 3: KNOWLEDGE TRANSFER

Knowledge is transferred from a research organization to industry through two primary routes. The first is the transfer of intellectual property. Research findings carried out at universities, colleges, non-profit organizations, and DOE and other federal laboratories are disseminated as published articles in peer-reviewed journals and in trade journals, presentations at national and international conferences, electronic media, or their organization's web site, and if exciting enough, via news conferences and press releases assuming the new results are not patentable. If, however, the research has some potential commercial value, this new information/data should be made available as soon as feasibly possible after filing a patent disclosure. However, before the patent is filed one could disseminate the results to companies that might be interested by contacting them directly to see: (1) if they are interested, (2) if they would sign a non-disclosure agreement, and (3) if they answer yes to both (1) and (2) then the information could be disclosed to them. However, all the companies must be treated equally and fairly.

The second route is highly effective when the research organization is connected with a university. This is exemplified by Ames Laboratory and Iowa State University. AL employs a significant number of ISU students in part time positions either as graduate research assistants or undergraduate research helpers. These science and engineering students, particularly at the bachelors and masters levels, transfer the skills and process the knowledge gained in working in the laboratory to their employers after they graduate.

QUESTION 4. U.S. RESEARCH ON RARE EARTH MINERALS

Rare earth research in the USA on mineral extraction, rare earth separation, processing of the oxides into metallic alloys and other useful forms (i.e. chlorides, carbonates, ferrites), substitution, and recycling is virtually zero. As is well-known, research primarily follows money; but prestige and accolades are other drivers; or when someone serendipitously comes up with an exciting idea for a research project. The lack of money and excitement accounts for the low level of research on the above topics.

Today some work on rare earth and actinide separation chemistry is directed toward treating waste nuclear products and environmental clean-up of radioactive materials in soils is being carried out at various DOE laboratories. This research may be beneficial to improving rare earth separation processes on a commercial scale.

Some research at various universities might be considered to be useful in finding substitutes for a given rare earth element in a high tech application. But generally the particular rare earth's properties are so unique it is difficult to find another element (rare earth or non-rare earth) as a substitute.

The Chinese have two large research laboratories which have significant research and development activities devoted to the above topics. They are the General Research Institute for Nonferrous Metals (GRINM) in Beijing, and the Baotou Research Institute of Rare Earths (BRIRE) in Baotou, Inner Mongolia. GRINM is a much larger organization than the Baotou group, but the rare earths activity is smaller than what is carried out at BRIRE. The Baotou Research Institute of Rare Earths is the largest rare earth research group in the world. Baotou is located about 120 miles from the large rare earth deposit in Inner

Mongolia and is the closest large city to the mine. This is the reason why BRIRE is located in Baotou.

Karl A. Gschneidner, Jr.
Ames Laboratory, U.S. Department of Energy and Department of Materials Science and Engineering Iowa State University Ames, IA 50011-3020 Phone: 515-294-7931; E-mail: cagey@ameslab.gov

Karl A. Gschneidner, Jr. was born on November 16, 1930 in Detroit, Michigan, and received his early education at St. Margaret Mary grade school and St. Bernard high school. He attended the University of Detroit, 1948-1952 and graduated with B.S. in Chemistry. He went to graduate school at Iowa State College (became Iowa State University in 1959) and in 1957 obtained a Ph.D. degree in Physical Chemistry studying under Distinguished Professor Frank H. Spedding and Professor Adrian H. Daane. He then worked in the plutonium research group at the Los Alamos Scientific Laboratory from 1957 through 1963. In 1963 he joined the Department of Metallurgy as an Associate Professor, and jointly as a group leader at the Ames Laboratory of Iowa State University. He was promoted to a full professor in 1967, and named a Distinguished Professor in 1979. In 1966 he founded the Rare-earth Information Center and served as its Director for 30 years. He was also the Program Director for Metallurgy and Ceramics at the Ames Laboratory from 1974 to 1979. He taught mostly graduate level courses, including x-ray crystallography, the physical metallurgy of rare earths, and alloying theory.

Gschneidner, sometimes known as "Mr. Rare Earths", is one of the world's foremost authorities in the physical metallurgy, and the thermal, magnetic and electrical behaviors of rare earth materials, a group of chemically similar metals naturally occurring in the earth's crust. His work lately has taken him into the field of magnetic refrigeration, a developing technology that has the potential for significant energy savings with fewer environmental problems than existing refrigeration systems.

Gschneidner has over 450 refereed journal publications and nearly 300 presentations to leading scientific gatherings worldwide to his credit. Holder of more than a dozen patents, he has been honored with numerous awards by governmental, professional, and industrial bodies, including recognition for his Ames Lab team's research in magnetic refrigeration by the U.S. Department of Energy in 1997 and with an Innovative Housing Technology Award in 2003.

In addition to the National Academy of Engineering, Gschneidner is also a Fellow of the American Society for Materials-International, The Minerals,

Metals and Materials Society, and the American Physical Society. In 2005, he was honored for 53 years of outstanding contributions to his field with a symposium at Iowa State that was attended by some of the world's leading experts in rare earth materials, many of them his former students or collaborators. He maintains an active research program with Ames Laboratory.

ACKNOWLEDGMENTS

The author wishes to thank his colleagues and associates for assisting him in putting this report together. They are: A.H. King, Director, Ames Laboratory; and K.A. Ament, I.E. Anderson, J.D. Corbett, D.L. Covey, K.B. Gibson, B.N. Harmon, S.L. Joiner, S.R. Karsjen, L.L. Jones, T.A. Lograsso, R.W. McCallum, V.K. Pecharsky, F.A. Schmidt, and C.J. Smith.

REFERENCES

[1]   Svec, HJ. 1-31 *"Prologue"* in *Handbook on the Physics and Chemistry of Rare Earths*, vol. 11, KA; Gschneidner, Jr. L. Eyring, Eds., Elsevier Science Publishers, 1988.

[2]   Goldman, JA. "National Science in the Nation's Heartland. The Ames Laboratory and Iowa State University, 1942-1965", *Technology and Culture*, 2000, 41, 435-459.

[3]   Goldman, JA. "Mobilizing Science in the Heartland: Iowa State College, the State University of Iowa and National Science during World War II", *The Annals of Iowa*, Fall 2000, 59, 374- 397.

[4]   Goldman, JA. "Frank Spedding and the Ames Laboratory: The Development of a Science Manager", *The Annals of Iowa*, Winter 2008, 67, 51-81.

[5]   Corbett, JD. "Frank Harold Spedding 1902-1984", *Biographical Memories*, 2001, 80, 1-28, The National Academy Press, Washington, DC.

[6]   Karsjen, SR. *"The Ames Project: History of the Ames Laboratory's Contributions to the Historic Manhattan Project*, 1942-1946", published by Ames Laboratory Public Affairs, Iowa State University, Ames, Iowa, 2003.

[7]   Gschneidner, KA; Jr. Pecharsky, VK. "Thirty Years of Near Room Temperature Magnetic Cooling: Where we are Today and Future Prospects", *Intern. J. Refrig*, 2008, 31, 945-961.

End Notes

[a] Coined by Chemical and Engineering News in the 1970s.
[b] At that time ISU was known as Iowa State College, but I will use ISU in this presentation. The name change officially occurred in 1959.
[c] Other team members were K. Dennis, M. Kramer and Dan Branagan, who moved to DOE's INEEL laboratory.

In: Rare Earth Minerals: Policies and Issues   ISBN: 978-1-61122-310-1
Editor: Steven M. Franks   ©2011 Nova Science Publishers, Inc.

Chapter 7

TESTIMONY OF STEVEN J. DUCLOS, CHIEF SCIENTIST AND MANAGER, MATERIAL SUSTAINABILITY, GE GLOBAL RESEARCH, BEFORE THE SUBCOMMITTEE ON INVESTIGATIONS AND OVERSIGHT

Steven J. Duclos[*]

INTRODUCTION

Chairman Miller and members of the Committee, it is a privilege to share with you GE's thoughts on how we manage shortages of precious materials and commodities critical to our manufacturing operations and what steps the Federal government can take to help industry minimize the risks and issues associated with these shortages.

---

[*] Corresponding author: Email: duclos@research.ge.com

BACKGROUND

GE is a diversified global infrastructure, finance, and media company that provides a wide array of products to meet the world's essential needs. From energy and water to transportation and healthcare, we are driving advanced technology and product solutions in key industries central to providing a cleaner, more sustainable future for our nation and the world.

At the core of every GE product are the materials that make up that product. To put GE's material usage in perspective, we use at least 70 of the first 83 elements listed in the Periodic Table of Elements. In actual dollars, we spend $40 billion annually on materials. 10% of this is for the direct purchase of metals and alloys. In the specific case of the rare earth elements, we use these elements in our Healthcare, Lighting, Energy, Motors, and Transportation products.

Nowhere in the company is our understanding of materials more evident than at GE Global Research, the hub of technology development for all of GE's businesses. Located just outside of Albany NY, GE scientists and engineers have been responsible for major material breakthroughs throughout our 110-year history. One of GE's earliest research pioneers, William Coolidge, discovered a new filament material, based on ductile tungsten, in 1909, which enabled us to bring the light bulb to every home. Just four years later, he developed a safe x-ray tube design for medical imaging. In 1953, GE scientist Daniel Fox developed LEXAN plastic, which is used in today's CDs and DVDs. It was even used in the helmets that US astronauts Neil Armstrong and Buzz Aldrin wore when they walked on the moon. More recently, GE scientists created a unique scintillator material, called Gemstone, which is the key component in GE Healthcare's newest High-Definition Computed Tomography (CT) medical imaging scanner that enables faster and higher resolution imaging.

Because materials are so fundamental to everything we do as a company, we are constantly watching, evaluating, and anticipating supply changes with respect to materials that are vital to GE's business interests. On the proactive side, we invest a great deal of time and resources to develop new materials and processes that help reduce our dependence on any given material and increase our flexibility in product design choices.

We have more than 35,000 scientists and engineers working for GE in the US and around the globe, with extensive expertise in materials development, system design, and manufacturing. As Chief Scientist and Manager of Material Sustainability at GE Global Research, it's my job to understand the latest trends in materials and to help identify and support new R&D projects with our businesses to manage our needs in a sustainable way.

Chairman Miller, I commend you for convening this hearing to discuss an issue that is vital to the future well being of US manufacturing. Without development of new supplies and more focused research in materials and manufacturing, such supply challenges could seriously undermine efforts to meet the nation's future needs in energy, healthcare, and transportation. What I would like to do now is share with you GE's strategy to address its materials needs, as well as outline a series of recommendations and indeed, a framework, for how the Federal government can strengthen its support of academia, government, and industry in this area.

COMMENTS AND RECOMMENDATIONS

The process that GE uses to evaluate the risks associated with material shortages is a modification of an assessment tool developed by the National Research Council in 2008. Risks are quantified element by element in two categories: "Price and Supply Risk", and "Impact of a Restricted Supply on GE". Those elements deemed to have high risk in both categories are identified as materials needing further study and a detailed plan to mitigate supply risks. The "Price and Supply Risk" category includes an assessment of demand and supply dynamics, price volatility, geopolitics, and co-production. Here we extensively use data from the US Geological Survey's Minerals Information Team, as well as in-house knowledge of supply dynamics and current and future uses of the element. The "Impact to GE" category includes an assessment of our volume of usage compared to the world supply, criticality to products, and impact on revenue of products containing the element. While we find this approach adequate at present, we are working with researchers at Yale University who are in the process of developing a more rigorous methodology for assessing the criticality of metals. Through these collaborations, we anticipate being able to predict with much greater confidence the level of criticality of particular elements for GE's uses.

Once an element is identified as high risk, a comprehensive strategy is developed to reduce this risk. Such a strategy can include improvements in the supply chain, improvements in manufacturing efficiency, as well as research and development into new materials and recycling opportunities. Often, a combination of several of these may need to be implemented.

Improvements in the global supply chain can involve the development of alternate sources, as well as the development of long-term supply agreements that allow suppliers a better understanding of our future needs. In addition, for elements that are environmentally stable, we can inventory materials in order to mitigate short-term supply issues.

Improvements in manufacturing technologies can also be developed. In many cases where a manufacturing process was designed during a time when the availability of a raw material was not a concern, alternate processes can be developed and implemented that greatly improve its material utilization. Development of near-net-shape manufacturing technologies and implementation of recycling programs to recover waste materials from a manufacturing line are two examples of improvements than can be made in material utilization.

An optimal solution is to develop technology that either greatly reduces the use of the at-risk element or eliminates the need for the element altogether. While there are cases where the properties imparted by the element are uniquely suitable to a particular application, I can cite many examples where GE has been able to invent alternate materials, or use already existing alternate materials to greatly minimize our risk. At times this may require a redesign of the system utilizing the material to compensate for the modified properties of the substitute material. Let's look at a few illustrative recent examples.

The first involves Helium-3, a gaseous isotope of Helium used by GE Energy's Reuter Stokes business in building neutron sensors for detecting special nuclear materials at the nation's ports and borders. The supply of Helium-3 has been diminishing since 2001 due to a simultaneous increase in need for neutron detection for security, and reduced availability as Helium-3 production has dwindled. GE has addressed this problem in two ways. The first was to develop the capability to recover, purify and reuse the Helium-3 from detectors removed from decommissioned equipment. The second was the accelerated development of Boron-10 based detectors that eliminate the need for Helium-3 in Radiation Portal Monitors. DNDO and the Pacific Northwest National Lab are currently evaluating these new detectors.

A second example involves Rhenium, an element used at several percent in super alloys for high efficiency aircraft engines and electricity generating turbines. Faced with a six-fold price increase during a three-year stretch from 2005 to 2008 and concerns that its supply would limit our ability to produce our engines, GE embarked on multi-year research programs to develop the capability of recycling manufacturing scrap and end-of-life components. A significant materials development effort was also undertaken to develop and certify new alloys that require only one-half the amount of Rhenium, as well as no Rhenium at all. This development leveraged past research and development programs supported by DARPA, the Air Force, the Navy, and NASA. The Department of Defense supported qualification of our reduced Rhenium engine components for their applications.

By developing alternate materials, we created greater design flexibility that can be critical to overcoming material availability constraints. But pursuing this path is not easy and presents significant challenges that need to be addressed. Because the materials development and certification process takes several years, executing these solutions requires advanced warning of impending problems. For this reason, having shorter term sourcing and manufacturing solutions is critical in order to "buy time" for the longer term solutions to come to fruition. In addition, such material development projects tend to be higher risk and require risk mitigation strategies and parallel paths. The Federal Government can help by enabling public-private collaborations that provide both the materials understanding and the resources to attempt higher risk approaches. Both are required to increase our chances of success in minimizing the use of a given element.

Another approach to minimizing the use of an element over the long term is to assure that as much life as possible is obtained from the parts and systems that contain these materials. Designing in serviceability of such parts reduces the need for additional material for replacement parts. The basic understanding of life-limiting materials degradation mechanisms can be critical to extending the useful life of parts, particularly those exposed to extreme conditions. It is these parts that tend to be made of the most sophisticated materials, often times containing scarce raw materials.

A complete solution often requires a reassessment of the entire system that uses a raw material that is at risk. Often, more than one technology can address a customer's need. Each of these technologies will use a certain subset of the periodic table – and the solution to the raw material constraint may involve using a new or alternate technology. Efficient lighting systems provide an excellent example of this type of approach. Linear fluorescent lamps use several rare earth elements. In fact, they are one of the largest consumers of Terbium, a rare earth element that along with Dysprosium is also used to improve the performance of high- strength permanent magnets. Light emitting diodes (LEDs), a new lighting technology whose development is being supported by the Department of Energy, uses roughly one-hundredth the amount of rare earth material per unit of luminosity, and no Terbium. Organic light emitting diodes (OLEDs), an even more advanced lighting technology, promises to use no rare earth elements at all. In order to "buy time" for the LED and OLED technologies to mature, optimization of rare earth usage in current fluorescent lamps can also be considered. This example shows how a systems approach can minimize the risk of raw materials constraints.

In addition to high efficiency lighting, GE uses rare earth elements in our medical imaging systems and in wind turbine generators. Rare earth permanent magnets are a key technology in high power density motors. These motors are vital to the nation's vision for the electrification of transportation, including automobiles, aircraft, locomotives, and large off-road vehicles. The anticipated growth in the use of permanent magnets and other rare earth based materials for efficient energy technologies mandates that we develop a broad base solution to possible raw material shortages. These solutions require the development of the sourcing, manufacturing efficiency, recycling, and material substitution approaches outlined above.

Based on our past experience I would like to emphasize the following aspects that are important to consider when addressing material constraints:

1. Early identification of the issue – technical development of a complete solution can be hampered by not having the time required to develop some of the longer term solutions.

2. Material understanding is critical – with a focus on those elements identified as being at risk, the understanding of materials and chemical sciences enable acceleration of the most complete solutions around substitution. Focused research on viable approaches to substitution and usage minimization greatly increases the suite of options from which solutions can be selected.

3. Each element is different and some problems are easier to solve than others – typically a unique solution will be needed for each element and each use of that element. While basic understanding provides a foundation from which solutions can be developed, it is important that each solution be compatible with real life manufacturing and system design. A specific elemental restriction can be easier to solve if it involves few applications and has a greater flexibility of supply. Future raw materials issues will likely have increased complexity as they become based on global shortages of minerals that are more broadly used throughout society.

Given increasing challenges around the sustainability of materials, it will be critical for the Federal government to strengthen its support of efforts to minimize the risks and issues associated with material shortages. Based on the discussion above, we make the following recommendations for the Federal government:

1. Appoint a lead agency with ownership of early assessment and authority to fund solutions — given the need for early identification of future issues, we recommend that the government enhance its ability to monitor and assess industrial materials supply, both short term and long term, as well as coordinate a response to identified issues. Collaborative efforts between academia, government laboratories, and industry will help ensure that manufacturing compatible solutions are available to industry in time to avert disruptions in US manufacturing.

2. Sustained funding for research focusing on material substitutions — Federal government support of materials research will be critical to laying the foundation upon which solutions are developed when materials supplies become strained. These complex problems will require collaborative involvement of academic and government laboratories with direct involvement of industry to ensure solutions are manufacturable.

3. With global economic growth resulting in increased pressure on material stocks, along with increased complexity of the needed resolutions, it is imperative that the solutions discussed in this testimony: recycling technologies, development of alternate materials, new systems solutions, and manufacturing efficiency have sustained support. This will require investment in long-term and high-risk research and development — and the Federal government's support of these will be of increasing criticality as material usage grows globally.

CONCLUSION

In closing, we believe that a more coordinated approach and sustained level of investment from the Federal government in materials science and manufacturing technologies is required to accelerate new material breakthroughs that provide businesses with more flexibility and make us less vulnerable to material shortages. Chairman Miller and members of the committee, thank you for your time and the opportunity to provide our comments and recommendations.

Dr. Steven J. Duclos

Steven Duclos is a Chief Scientist at the General Electric Global Research Center in Niskayuna, New York, and manages GE's Material Sustainability Initiative. The Material Sustainability initiative addresses GE's risks in the availability and sustainability of the company's raw material supply, by developing technologies that reduce the use, support the recycling, and enable substitution of lower-risk materials.

From 2000 to 2008 Dr. Duclos managed the Optical Materials Laboratory, also at GE GRC. The laboratory is responsible for development of advanced materials for a broad spectrum of GE businesses, including its Lighting and Healthcare businesses. From 1994 to 2004 Dr. Duclos served on the Executive Committee of the New York State Section of the American Physical Society. Prior to joining the GE Global Research Center in 1991 he was a post-doc at AT&T Bell Laboratories in Murray Hill, New Jersey.

Dr. Duclos received his B.S. degree in Physics in 1984 from Washington University in St. Louis, M.S. degree in Physics from Cornell University in Ithaca, New York in 1987, and Ph.D. in Physics from Cornell in 1990. He is the recipient of an AT&T Bell Laboratories Pre-doctoral Fellowship and the 1997 Albert W. Hull Award, GE Global Research's highest award for early career achievement.

In: Rare Earth Minerals: Policies and Issues   ISBN: 978-1-61122-310-1
Editor: Steven M. Franks   ©2011 Nova Science Publishers, Inc.

Chapter 8

WRITTEN TESTIMONY OF MARK A. SMITH, CHIEF EXECUTIVE OFFICER, MOLYCORP MINERALS, LLC, BEFORE THE SUBCOMMITTEE ON INVESTIGATIONS AND OVERSIGHT, HEARING ON "RARE EARTH MINERALS AND 21ST CENTURY INDUSTRY"

Mark A. Smith

INTRODUCTION

Chairman Miller, Ranking Member Broun, and Members of the Subcommittee, I want to thank you for the opportunity to share my observations, experiences, and insights on the subject of rare earths, the critical role they play in the technologies that will shape our future, the looming supply challenges that are ahead of us, and the work we are doing at our facility at Mountain Pass, California. This is the first Committee to hold a hearing specifically on this important topic, and I want to commend you for your leadership and forethought.

I'm the CEO of rare earths technology company Molycorp Minerals, LLC. Molycorp owns the rare earth mine and processing facility at Mountain Pass, California, one of the richest rare earth deposits in the world, and we are the only active producer of rare earths in the Western Hemisphere. I have worked with Molycorp and its former parent companies, Unocal and Chevron, for over 25 years, and have watched closely the evolution of this industry over the past decade. It has been remarkable to watch the applications for rare earths explode. However, as rare earth-based technologies have become more and more essential, the U.S., which invented rare earth processing and manufacturing technology, has become almost completely dependent on China for access to rare earths and, more specifically, the metals, alloys and magnets that derive from them.

On its face it may not seem any more disconcerting than any other material dependency. However, it is the combination of three key factors that make this situation one of urgent concern to policymakers: 1) the indispensability of rare earths in key clean energy and defense technologies; 2) the dominance of rare earth production by one country, China, and 3) China's accelerating consumption of their own rare earth resources, leaving the rest of the world without a viable alternative source.

The development of clean energy technology is a top national priority, as these innovations are key to our broader national objectives of greater energy security and independence, reduced carbon emissions, long term economic competitiveness, and robust job creation. Yet all of these crucial national objectives become less achievable if we lack access to rare earth resources.

Our company has produced rare earths for 57 years, and we are in the process of restarting active mining and down-stream processing at Mountain Pass. We are redeveloping our facilities to dramatically increase our production, and we are executing a strategy to rebuild the rare earth metal and magnet manufacturing capabilities that our country has lost in the past decade. This effort will help to address rare earth access concerns and may help to catalyze clean tech manufacturing, but the lingering question is how quickly we can make this happen, as the looming supply concerns seem to accelerate every day.

Below I offer my perspective on rare earths and their applications, America's rare earth capability gap, the global supply concerns and their implications, our work at Molycorp to expand our domestic rare earth access, and the role the federal government can play to help address the looming supply concerns.

RARE EARTH ELEMENTS AND KEY APPLICATIONS

Rare earths are a group of 17 elements (atomic numbers 57-7 1 along with Sc and Y) whose unique properties make them indispensible in a wide variety of advanced technologies. One rare earth in particular – neodymium (Nd) – is used to create the very high powered but lightweight magnets that have enabled miniaturization of a long list of consumer electronics, such as hard disk drives and cell phones. While high-tech applications such as these have dominated the usage of rare earths over the past decade, it is their application in clean energy technologies and defense systems that has brought heightened attention to rare earths.

Rare earths are indispensable in a wide variety of clean energy technologies. Rare earth metals are used in the advanced nickel-metal hydride (NiMH) batteries that are found in most current model hybrids; powerful rare earth magnets enable next generation wind turbines, electric vehicle motors, and hybrid vehicle motors and generators; and rare earth phosphors are what illuminate compact fluorescent light bulbs. On the defense side, missile guidance systems, military electronics, communications and surveillance equipment all require rare earths. None of these technologies will work without rare earths, and yet each of these technologies is tied closely to some of the nation's highest national priorities, our energy and national security.

The list of rare earth applications is long and varied, but there are additional applications that are worth noting specifically. The automotive sector is a big user of rare earths. Cerium is used to polish glass and provides protection from UV rays. In the 1970s, rare earths replaced palladium for use in catalytic converters, and if palladium were still used today, cars would be significantly more expensive. They are also used in petroleum refining and as diesel additives.

At Molycorp, we have also found a use for cerium in water filtration. We have developed proprietary water filtration technology that has applications in industrial wastewater treatment, clean water production in the developing world, and the recreation and backpacking market.

The diagram below offers a broader view of rare earths' applications:

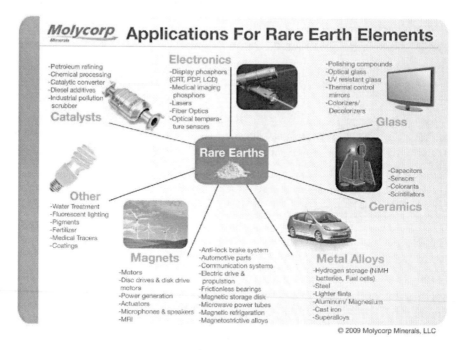

Despite their name, rare earths are not rare. If you were to go outside right now and grab a handful of dirt from the ground, it would contain rare earths. However, it is far more difficult to find rare earths in a concentration high enough to be mined and separated economically. When rare earths are extracted from the ground, the ore contains all of the rare earths, and it is through highly complex separation processes that each individual rare earth oxide can be produced. It is this separation process that largely drives the cost of rare earth production. Ore bodies that contain rare earths at percentages in the low single digits cannot be mined economically under current prices for rare earths.

Thus, today, there are only 3 known and verified locations where a sufficiently high concentration of rare earths exists: Baotou, China; Mountain Pass, California, where Molycorp's mine is located; and Mt. Weld, Australia, which has a rich ore deposit but none of the infrastructure necessary to begin extraction, separation, and distribution to market. Given these circumstances, Molycorp's mine at Mountain Pass is clearly one of the only rare earth resources in the world that is immediately minable, economically viable, and can provide a near-term source of rare earth materials. With supply concerns becoming increasingly imminent, the greatest challenge facing Molycorp is the speed at which we can bring these needed resources online. I will discuss this in further detail later in this testimony.

INDUSTRIAL SUPPLY CHAIN AND AMERICA'S CAPABILITY GAP

One of the biggest challenges in raising awareness and understanding about rare earths is that they are found so early in the industrial supply chain that it is difficult to contemplate their usage in products that we see every day. To illustrate this point, consider the example of the new generation of wind turbines, which employ rare earth-based permanent magnet generators with reliability and efficiency improvements of 70% over the current industry standard. Below is a simplified supply chain:

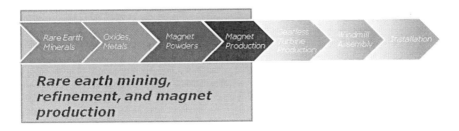

Once the rare earths are mined, they are separated and converted to oxides and then converted into metals. The metals are then manufactured into alloys and magnet powders. The powders are then bonded or sintered to form the magnets required for turbine production. The turbine, in turn, is included in the windmill assembly, and the final product is installed. All of the functions within the green box are necessary to be able to produce the magnets required for this clean energy technology and so many others. However, other than the rare earth mineral extraction and conversion to oxides, the other manufacturing capabilities in the green box no longer exist in the United States. The U.S. did at one time possess all of these capabilities, and in fact, these technologies largely originated here. However, over the past decade as American manufacturing has steadily eroded, the U.S. has ceded this technological ground to competitors in China, Japan and Germany.

China has become particularly dominant, and some would contend that it has been by design. In the early 1990s, China's Deng Xiaoping was quoted as saying, "There is oil in the Middle East; there is rare earth in China."[1] China realized that it had a significant natural resource advantage, and through the development of new applications in an ever-expanding number of advanced technologies, China helped to grow the market for rare earths from 40,000 tons in the early 1990's to roughly 125,000 tons in 2008. It is over that same period that, due to a variety of factors, the U.S. ceased active mining of rare earths.

While the U.S. still possesses the technical expertise, we have lost the necessary infrastructure to manufacture the rare earth metals and magnets that fuel next generation technologies. The last rare earth magnet manufacturer in the U.S. was a company called Magnaquench, formerly located in Valparaiso, Indiana, and owned by General Motors. Magnaquench and all of its U.S. assets were sold to a Chinese company in the early 2000s in an effort to help GM gain access to the Chinese market.[2] Two domestic companies can produce small quantities of rare earth based alloys but none can convert the rare earth oxides to metal. The result is a significant rare earth "capability gap" in the U.S. that has the potential to quickly become a major strategic and economic disadvantage.

GLOBAL SUPPLY CONCERNS AND IMPLICATIONS FOR THE U.S.

Today, the production of rare earths, and the metals and magnets that derive from them, is overwhelmingly dominated by China. At present, China produces 97% of the world's rare earth supply, almost 100% of the associated metal production, and 80% of the rare earth magnets. Complicating this picture even further, China's national consumption of rare earth resources is growing at an intense pace, consistent with their meteoric GDP growth, and it is leaving the rest of the world with less of these critical materials just as the clean energy economy is beginning to gain momentum. As the chart below from rare earths research firm, the Industrial Minerals Company of Australia (IMCOA) demonstrates, China's massive production has been able to satisfy both their own internal needs and those of the rest of the world. However, as the blue line indicates Chinese demand for its own rare earths will soon match, if not eclipse, its own internal supply, and with global demand (in yellow) growing at a parallel pace, there is a significant production gap – around 60,000 tons – that must be filled in a very short timeframe.

IMCOA's previous forecasts concluded that this critical shortage for the rest of the world outside of China would occur by 2012, but China has recently said that it intends to be the world's largest producer of wind energy and electric vehicles and has committed $150 billion and $29 billion to these two respective clean technology sectors (by comparison, the entire amount of stimulus funding under the American Recovery and Reinvestment Act directed at all areas of clean energy deployment was $60 billion). The new, more efficient wind turbines that use rare earth permanent magnet generators require around 2 tons of rare earth magnets per windmill. The rare earth industry has never seen this level of demand. To date, rare earth producers like Molycorp have filled orders by the pound or kilogram, not by the ton. If China's commitment holds true, this will vastly accelerate their consumption of rare earths and speed up the date when the rest of the world will find its access to rare earths severely limited.

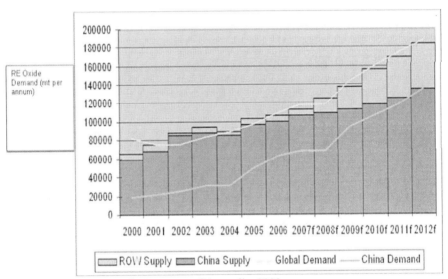

Source: Industrial Minerals Company of Australia, 2008

Around the same time that China was outlining its clean energy investments, it also began to consider steps to reduce the availability of its rare earths to the rest of the world. As if the demand forecasts weren't disconcerting enough, China heightened international supply concerns last fall when its Interior Ministry signaled that it would further restrict its exports of rare earth resources. China has been steadily decreasing its exports by an average of 6% per year since 2002, but these new restrictions portend a more aggressive effort to use its own resources domestically. This critical issue was featured on the front page of the New York Time's business section on August 31, 2009, and I've included the article for the Committee's record.

Finally, in late December, China announced that it will begin to stockpile rare earths. It is our estimation that if they are announcing it officially to the rest of the world, it is highly likely that the stockpiling has been occurring for some time. Regardless, it will have a further depressive effect on global supply.

ENERGY SECURITY AND GLOBAL COMPETITIVENESS

Disruption in the global supply of rare earths poses a significant concern for America's energy security and clean energy objectives, its future defense needs, and its long-term global competitiveness. Rare earths may not be familiar to most people, hidden deep in the industrial supply chain, but they are absolutely indispensable for so many of the advanced technologies that will allow us to achieve critical national objectives.

Efforts to decrease U.S. dependence on foreign oil and develop a clean energy economy, as well as the jobs that come with it, have received broad bipartisan support, and few would disagree that the U.S. must diversify its sources of energy and slow the demand for fossil fuels. Wind power and electric vehicles (EVs) have emerged as technologies that will play important roles in these efforts, and the U.S. has indicated it intends to be a leader in both. As noted above, the most efficient wind turbines require multiple tons of rare earths, and as the U.S. moves to increase the percentage of power that comes from wind, there will be a commensurate increase in domestic demand for rare earths. The American automotive industry is expanding the number of hybrid, plug-in hybrid (PHEV), and full electric vehicle (EV) models in an effort to produce far more fuel efficient products, and yet many of the advanced batteries that power hybrids and PHEVs utilize several kilograms of rare earth metals in each unit. The motors and generators in these new vehicles also use several kilograms of rare earth permanent magnets. Similar implications exist for our national defense capabilities. From military communications equipment to missile guidance systems, rare earths enable a long list of advanced defense technologies. We have had extensive discussions with the Department of Defense (DoD), and they are paying far greater attention to this concern. In fact, the FY2011 DoD Authorization signed into law last October included a provision requiring the Department to submit a report to Congress no later than April 1, 2010, assessing the usage of rare earth materials in DoD's supply chain, looking at projected availability for use by DoD, the extent to which the DoD is dependent on rare earth materials, steps that the Department is taking to address any risks to national security, and recommendations for further action.

Access to rare earths is obviously essential, but without rebuilding the manufacturing capacity to produce rare earth metals and magnets, the U.S. could find itself dependent on China for key technological building blocks. But even this scenario presumes that the U.S. has the manufacturing capabilities to put Chinese rare earth materials to use in final products. Right now, given the current state of U.S. manufacturing, it is unfortunately more likely that we would become a raw material supplier to Chinese manufacturers.

Viewed through this lens, the domestic development of rare earth resources and manufacturing capabilities is not only a strategic necessity but also a potential catalyst for job growth in the clean energy and advanced technology manufacturing sectors. If these resources and capabilities were built up domestically, it could have a multiplier effect on downstream, value added manufacturing. Consider China's experience. It has to create 10-15 million jobs a year just to accommodate new entrants into its job market, and it has viewed the rare earths industry as a "magnet" for jobs. China repeatedly attracted high-tech manufacturers to move to its shores in exchange for access to rare earths among other enticements. The U.S. could experience a similar jobs boost by making a concerted effort to rebuild the clean energy supply chain, beginning with rare earths, within its borders.

MOLYCORP MINERALS' MINING TO MAGNETS STRATEGY

Molycorp Minerals has been in the rare earths business for 57 years, and while the company and its facilities have changed ownership over the years, it has remained one of the world's only viable sources of rare earth minerals. On October 1, 2008, a group of U.S. based investors, including myself, formed Molycorp Minerals, LLC, and we acquired from Chevron its rare earth assets at Mountain Pass, which the U.S. Geological Survey has deemed "the greatest concentration of rare earth minerals now known." From the outset, we have sought to combine this world-class rare earth deposit with a "mining to magnets" strategy. Our redevelopment of Mountain Pass is the starting point of a broader effort to reestablish domestic manufacturing of the rare earth metals, alloys and magnets that enable and are indispensible to the clean energy economy and advanced technology manufacturing.

Our work at Mountain Pass provides a timely, well-planned, and economically viable means to address the rare earth access challenges on the shortest timeline possible. While Molycorp has been processing existing rare earth stockpiles since 2007, it has invested $20 million to begin the restart of active mining. Our team matches this remarkable natural resource with 57 years of rare earth mining, milling, and processing experience. We have obtained the necessary 30-year mine plan permit, and the Environmental Impact Report for the mining-tooxides portion of the redevelopment has been reviewed and approved by all applicable federal, state, and local agencies. Molycorp's footprint will be limited to its privately-held land, using state-of-the-art technologies for water treatment and mineral recovery. Through new advances in our production processes, Molycorp will produce 20,000 tons, or 40,000,000 pounds, of rare earth oxides per year, and our increased production and capabilities can potentially create 900 new jobs for the hard hit San Bernardino-Riverside and Henderson-Las Vegas regions of California and Nevada. Molycorp is the only domestic rare earth provider that stands "shovel- ready" to create jobs and commence the mining-to-magnets work required to meet multiple customer-specific product demands.

Access to the raw, rare earth minerals is obviously essential, but as noted earlier, it resolves only part of the challenge. As part of our mining to magnets development, we will build out the metals, alloying and magnet powder manufacturing capabilities. We would also establish the production of rare earth permanent magnets, all here in the United States. Our company is uniquely well-positioned to rebuild these early steps in the clean energy supply chain and fully extend the value and capabilities of the rare earth resources at Mountain Pass.

ENVIRONMENTAL STEWARDSHIP AS A SOURCE OF COST-COMPETITIVENESS

Many industry observers question how a U.S. producer of rare earths can ever compete with the Chinese, when the possibility always lingers that the Chinese could flood the market and dramatically depress rare earth prices, a practice they have demonstrated previously. We have spent the better part of the past 8 years developing the answer to this question. We changed the orientation of our thinking and discovered that by focusing principally on energy and resource efficiency, we could make major improvements in our cost competitiveness while at the same time advance our environmental stewardship.

We will incorporate a wide variety of manufacturing processes that are new to the rare earth industry, which will increase resource efficiency, improve environmental performance, and reduce carbon emissions. Specifically:

- Our overall processing improvements will almost cut in half the amount of raw ore needed to produce the same amount of rare earth oxides that we have produced historically. This effectively doubles the life of the ore body and further minimizes the mine's footprint;
- Our extraction improvements will increase the processing facility's rare earth recovery rates to 95% (up from 60-65%) and decrease the amount of reagents needed by over 30%;
- Our reagent recycling, through proprietary technology that Molycorp has developed, could lead to even greater decreases in reagent use;
- Our new water recycling and treatment processes reduce the mine's fresh water usage from 850 gallons per minute (gpm) to 30 gpm — a 96% reduction;
- Finally, the construction of a Combined Heat and Power (CHP) plant — fueled by natural gas — will eliminate usage of fuel oil and propane. This will significantly reduce the facility's carbon emissions, reduce electricity costs by 50%, and improve electricity reliability.

These process improvements fundamentally reverse the conventional wisdom that superior environmental stewardship increases production costs. At the same time, we significantly distinguish ourselves from the Chinese rare earth industry that has been plagued by a history of significant environmental degradation, one that it is just beginning to recognize and rectify.

NEED FOR FEDERAL LEADERSHIP

Over the past year, I have spent a significant amount of time in Washington meeting with Members of Congress and their staffs as well as officials in a variety of federal agencies to direct greater attention to this issue. I'm pleased to report, just over one year since we began our efforts, that the federal government is beginning to take meaningful steps toward understanding and addressing our rare earth vulnerabilities. The question remains, however, if it will be able to make its assessments, determine the required actions, and execute them within a timeline that seems to be accelerating daily.

In each of these meetings, and as this Committee has also inquired, I am asked what role the federal government should play in tackling this pressing concern, and I believe that there are 4 areas where it can have the greatest near- and long-term impact: 1) federally based financing and/or loan guarantee support for highly capital intensive projects like ours; 2) assistance rebuilding America's rare earth knowledge infrastructure (university-based rare earth research, development of academic curricula and fields of study, training and exposure to the chemical and physical science related to rare earths, etc.); 3) increased interagency collaboration at the highest levels on the impact of rare earth accessibility on major national objectives; and 4) funding competitive grants for public and private sector rare earth research. I'll explore each in greater depth below:

- **Financing support:** Given the size, scale, ambition, and necessity of Molycorp's redevelopment efforts, we submitted an application for the Department of Energy's Loan Guarantee Program (LGP). We believed that the program was well-suited for our project, particularly given that the project's substantial implications closely match the program's paramount objectives. Traditional bank financing in the current climate – with very short repayment periods and interest rates near double digits – is not economically feasible. The LGP offers longer term financing and lower interest rates and would allow Molycorp to accelerate development in the near-term while ensuring rare earth resource availability in the long term. However, the DOE summarily rejected our application in December, saying that the project did not qualify as a "New or Significantly Improved Technology." We reviewed the relevant portion of the Rule, Section 609.2, and our project meets every one of the stated criteria. We requested further discussion with the DOE to understand how it came to its conclusion and how Molycorp might proceed. After almost two months, the DOE finally responded to our request. During the meeting, the DOE contended that this project goes "too far upstream" and that the program was not intended to cover mining projects. We have yet to find the legislative or regulatory language that provides such a limitation. However, it appears we may need to ask Congress for legislative direction or possibly new legislative language specifically authorizing the use of loan guarantees for strategically important projects like this. Our frustrations with the loan guarantee notwithstanding, I still believe that this kind of financing support is exactly what a project like ours needs. We will be in a very strong position to both raise our portion of the capital to execute the project and repay the loan well-within the required timeline. We will continue to pursue this financing support despite the DOE's current position.

- **Rebuilding the rare earth knowledge infrastructure:** The United States used to be the world's preeminent source of rare earth information and expertise, but it has ceded that advantage over the past decade, as its position in the industry has become subordinate to China and other countries in East Asia. The federal government, and the House Science and Technology Committee in particular, can play a pivotal role in reestablishing that institutional knowledge and expertise and sharing it with a wider audience of researchers, scholars, and practitioners here in the U.S. and abroad. At Molycorp, we are fortunate to have a team of 17 rare earth researchers and technologists who are second to none in the world, but almost all of them had no previous expertise in rare earths prior to joining Molycorp. It will be difficult for the U.S. to reestablish its preeminence without a concerted effort to attract the brightest scientists and researchers to the field of rare earths. Rebuilding the knowledge infrastructure and the research support will go a long way toward that goal. Dr. Gerschneidner, who I'm honored to testify with today, is regarded as the father of rare earths, and his work at Ames Laboratory and Iowa State University as well as the great work being done by Dr. Eggert and his colleagues at the Colorado School of Mines can serve as the foundation on which to expand America's rare earth expertise. As a reminder of the rest of the world's interests and actions in this regard, the Korea Times recently reported that Korea is developing rare earth metals for industrial use at a government-funded research center.

- **Interagency Cooperation:** Over the past several months, we have been very pleased to learn about efforts within many federal agencies to direct specific attention to rare earth issues. We have been in direct contact with the Departments of Defense, Commerce, and State, and each is examining this issue within the unique context of their agencies' work. It is also worth noting that the Commerce Department convened a group of stakeholders from both the government and the private sector in December, 2009, which included representatives from DoD, GAO, USTR, and OSTP. We have also had multiple discussions with the Office of Science and Technology Policy directly and have been very appreciative of their engagement on this issue. In fact, OSTP, along with Commerce, is facilitating interagency collaboration going forward. While we are encouraged by these recent efforts, it is our hope that the agencies and the White House recognize that the global supply-demand challenges are approaching at an increasingly rapid pace and that their efforts should reflect the requisite urgency.

- **Funding support for rare earth research:** Part of China's success in growing and dominating the market for rare earths can be attributed to their efforts to find and commercialize new applications for rare earth materials. Federal funding support for competitive grants specifically directed at rare earth research will help to expand the U.S.'s ability to do the same. This has the potential to broaden the economic impact of rare earths, and contribute to the goal mentioned above of reestablishing America's superior expertise in rare earth research.

CONCLUSION

The global rare earth supply concerns facing the U.S. and all other countries outside China are obviously disconcerting, but they are not insurmountable. A combination of geologic good fortune and an accelerated effort to ramp up domestic production and rebuild lost manufacturing capabilities could provide a solution for the U.S. and ensure that our leading national objectives are not jeopardized. At Molycorp, our "mining to magnets" strategy is far more than an approach to a new business, it is a cause with far reaching implications. If executed effectively, it could prove to be catalytic for our development of a clean energy economy and the resurgence of domestic manufacturing. This project will have meaningful and significant impact on leading national priorities, and as such, we stand ready to work with Congress and the Administration to find ways to accelerate our work at Mountain Pass and bring these needed capabilities online as soon as possible.

Thank you once again for the opportunity to share my perspective on rare earths, and I look forward to working with the Committee in the weeks and months to come as it continues to examine this important topic and determine potential actions.

The New York Times

September 1, 2009; pp. B1, B4

Business

China Tightens Grip on Rare Minerals

By KEITH BRADSHER

HONG KONG - China is set to tighten its hammerlock on the market for some of the world's most obscure but valuable minerals.

China currently accounts for 93 percent of production of so-called rare earth elements - and more than 99 percent of the output for two of these elements, dysprosium and terbium, vital for a wide range of green energy technologies and military applications like missiles.

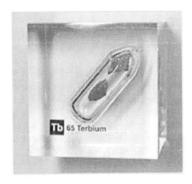

Juergen Bauer/www.smart-elements.com

China is cutting exports of elements like terbium, pressuring manufacturers to open factories there.

Deng Xiaoping once observed that the Mideast had oil, but China had rare earth elements. As the Organization of the Petroleum Exporting Countries has done with oil, China is now starting to flex its muscle.

Even tighter limits on production and exports, part of a plan from the Ministry of Industry and Information Technology, would ensure China has the supply for its own technological and economic needs, and force more manufacturers to make their wares here in order to have access to the minerals.

In each of the last three years, China has reduced the amount of rare earths that can be exported. This year's export quotas are on track to be the smallest yet. But what is really starting to alarm Western governments and multinationals alike is the possibility that exports will be further restricted.

Chinese officials will almost certainly be pressed to address the issue at a conference Thursday in Beijing. What they say could influence whether Australian regulators next week approve a deal by a Chinese company to acquire a majority stake in Australia's main rare-earth mine.

The detention of executives from the British-Australian mining giant Rio Tinto has already increased tensions.

They sell for up to $300 a kilogram, or up to about $150 a pound for material like terbium, which is in particularly short supply. Dysprosium is $110 a kilo, or about $50 a pound. Less scare rare earth like neodymium sells for only a fraction of that.

(They are considerably less expensive than precious metals because despite the names, they are found in much higher quantities and much greater concentrations than precious metal.)

China's Ministry of Industry and Information Technology has drafted a six-year plan for rare earth production and submitted it to the State Council, the equivalent of the cabinet, according to four mining industry officials who have discussed the plan with Chinese officials. A few, often contradictory, details of the plan have leaked out, but it appears to suggest tighter restrictions on exports, and strict curbs on environmentally damaging mines.

 Beijing officials are forcing global manufacturers to move factories to China by limiting the availability of rare earths outside China. "Rare earth usage in China will be increasingly greater than exports," said Zhang Peichen, the deputy director of the government-linked Baotou Rare Earth Research Institute.

Some of the minerals crucial to green technologies are extracted in China using methods that inflict serious damage on the local environment. China dominates global rare earth production partly because of its willingness until now to tolerate highly polluting, low-cost mining.

The ministry did not respond to repeated requests for comment in the last eight days. Jia Yinsong, a director general at the ministry, is to speak about China's intentions Thursday at the Minor Metals and Rare Earths 2009 conference in Beijing.

Until spring, it seemed that China's stranglehold on production of rare earths might weaken in the next three years - two Australian mines are opening with combined production equal to a quarter of global output.

But both companies developing mines - Lynas Corporation and smaller rival, Arafura Resources - lost their financing last winter because of the global financial crisis. Buyers deserted Lynas's planned bond issue and Arafura's initial public offering.

Rare Wealth

China accounts for the vast majority of the world's production of rare earths — 17 elements — which are used in a wide array of products.

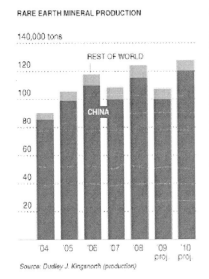

RARE EARTH MINERAL PRODUCTION

RARE EARTHS	ATOMIC NO.	COMMERICAL USE
Scandium	21	Stadium lights
Yttrium	39	Lasers
Lanthanum	57	Electric car batteries
Cerium	58	Lens polishes
Praseodymium	59	Searchlights, aircraft parts
Neodymium	60	High-strength magnets
Promethium	61	Portable X-ray units
Samarium	62	Glass
Europium	63	Compact fluorescent bulbs
Gadolinium	64	Neutron radiography
Terbium	65	High-strength magnets
Dysprosium	66	High-strength magnets
Holmium	67	Glass tint
Erbium	68	Metal alloys
Thulium	69	Lasers
Ytterbium	70	Stainless steel
Lutetium	71	None

Source: Dudley J. Kingsnorth (production)   THE NEW YORK TIMES

Mining companies wholly owned by the Chinese government swooped in last spring with the cash needed to finish the construction of both companies' mines and ore processing factories. The Chinese companies reached agreements to buy 51.7 percent of Lynas and 25 percent of Arafura.

The Arafura deal has already been approved by Australian regulators and is subject to final approval by shareholders on Sept. 17. The regulators have postponed twice a decision on Lynas, and now face a deadline of next Monday to act.

Matthew James, an executive vice president of Lynas, said that the company's would-be acquirer had agreed not to direct the day-to-day operations of the company, but would have four seats on an eight-member board.

Expectations of tightening Chinese restrictions have produced a surge in the last two weeks in the share prices of the few non-Chinese producers that are publicly traded. In addition to the two Australian mines, Avalon Rare Metals of Toronto is trying to open a mine in northwest Canada, and Molycorp Minerals is trying to reopen a mine in Mountain Pass, Calif.

Unocal used to own the Mountain Pass mine, which suspended mining in 2002 because of weak demand and a delay in an environmental review. State-owned CNOOC of China almost acquired the mine in 2005 with its unsuccessful bid for Unocal, which was bought instead by Chevron; Chinese buyers tried to persuade Chevron to sell the mine to them in 2007, but Chevron sold it to Molycorp Minerals, a private American group.

A single mine in Baotou, in China's Inner Mongolia, produces half of the world's rare earths. Much of the rest - particularly some of the rarest elements most needed for products from wind turbines to Prius cars - comes from small, often unlicensed mines in southern China.

China produces over 99 percent of dysprosium and terbium and 95 percent of neodymium. These are vital to many green energy technologies, including high-strength, lightweight magnets used in wind turbines, as well as military applications.

To get at the materials, powerful acid is pumped down bore holes. There it dissolves some of the rare earths, and the slurry is then pumped into leaky artificial ponds with earthen dams, according to mining specialists.

The Ministry of Industry and Information Technology has cut the country's target output from rare earth mines by 8.1 percent this year and is forcing mergers of mining companies in a bid to improve technical standards, according to the government-controlled China Mining Association, a government-led trade group.

General Motors and the United States Air Force played leading roles in the development of rare-earth magnets. The magnets are still used in the electric motors that control the guidance vanes on the sides of missiles, said Jack Lifton, a chemist who helped develop some of the early magnets.

But demand is surging now because of wind turbines and hybrid vehicles.

The electric motor in a Prius requires 2 to 4 pounds of neodymium, said Dudley Kingsnorth, a consultant in Perth, Australia, whose compilations of rare earth mining and trade are the industry's benchmark.

Mr. Lifton said that Toyota officials had expressed strong worry to him on Sunday about the availability of rare earths.

Toyota and General Motors, which plans to introduce the Chevrolet Volt next year with an electric motor that uses rare earths, both declined on Monday to comment.

Yuriko Nakao/Reuters

Toyota's Prius hybrids use several pounds of neodymium, a rare earth, in their electric motors.

Rick A. Lowden, a senior materials analyst at the Defense Department, told a Congressional subcommittee in July that his office was reviewing a growing number of questions about the availability of rare earths.

China is increasingly manufacturing high-performance electric motors, not just the magnets.

"The people who are making these products outside China are at a huge disadvantage, and that is why more and more of that manufacturing is moving to China," Mr. Kingsnorth said.

Correction: September 4, 2009

An article on Tuesday about China's tightening control over the production of rare earth minerals misidentified the country in which Avalon Rare Metals, a non-Chinese producer, was trying to open a new mine. It is Canada, in the northwest area, not Australia.

End Notes

[1] Cox, Clint. (2006, October 10). *Rare earth may be China's checkmate.* Retrieved from the Anchor House web site on February 5, 2010 at http://www.theanchorsite.com/2006/10/.

[2] Moberg, David. (2004, January 23). *Magnet Consolidation Threatens Both U.S. Jobs and Security.* Retrieved from the In These Times web site on February 5, 2010 at http://www.inthesetimes.com/article/685/.

In: Rare Earth Minerals: Policies and Issues   ISBN: 978-1-61122-310-1
Editor: Steven M. Franks   ©2011 Nova Science Publishers, Inc.

Chapter 9

TESTIMONY OF TERENCE P. STEWART, MANAGING PARTNER, LAW OFFICES OF STEWART AND STEWART, BEFORE THE SUBCOMMITTEE ON INVESTIGATIONS AND OVERSIGHT, HEARING ON "RARE EARTH MINERALS AND 21ST CENTURY INDUSTRY"

Terence P. Stewart

Mr. Chairman and members of the Subcommittee. Good afternoon. I am pleased to appear this afternoon as part of your hearing on rare earth minerals and 21st century industry to try to address three questions that I understand are of interest to the Subcommittee:

1. How do Chinese actions in the rare earths sector fit into China's policies on strategic industries and economic development?
2. Are there policies that th e Federal Government can adopt, or strategies that the U.S. private sector can adopt, that can help assure a consistent and sustainable domestic supply of economically and militarily critical materials such as rare earths?

3. Are there policies that the Federal Government can adopt, or strategies that the U.S. private sector can adopt, that would support firms dependent on rare earth elements to retain their manufacturing capacity in the U.S.?

Let me start with some acknowledgments on my limitations as a witness on rare earth minerals. First, my background and expertise is on international trade law matters, including the World Trade Organization, and manufacturing competitiveness issues. Others on the panel today are the experts on minerals in general or rare earth minerals policies.

Our firm, over the years, has looked at many aspects of the U.S.-China relationship and has prepared for the U.S.-China Economic and Security Review Commission various studies looking at the trade and manufacturing impacts of China's practices. For example, on March 24, 2009 I testified at a hearing before the Commission on "China's Industrial Policy and its Impact on U.S. Companies, Workers and the American Economy."

Rare earths are not part of the current WTO challenges brought by the U.S., the EU and Mexico of China's export restraints on various materials. (see WT/DS394/1, China – Measures Related to the Exportation of Various Raw Materials, request for consultations by the United States) Purchasers of rare earths are concerned about similar types of restraints imposed by China on rare earth minerals and that there have been discussions by the U.S. with key allies about a possible future case. *Inside U.S.-China Trade*, October 21, 2009, "U.S. and Allies Discuss Rare Earth Metals Action at WTO."

Now let me turn to the questions of interest to the Subcommittee.

CHINA'S ACTIONS ON RARE EARTH MINERALS

What China is doing on rare earth minerals mirrors what it is doing on a large number of other raw materials: reducing availability of supply for global customers and/or making purchases more expensive through the imposition of export duties, export licenses, etc. The objective can be to encourage foreign investors to move investment to China to produce downstream products in the Middle Kingdom versus overseas, or to ensure low priced supplies for sectors in China targeted for rapid industrial growth.

China's most recent five-year plan (covering 2006-20 10) continues to focus development in certain sectors and to ensure a leading role for state-owned enterprises ("SOEs") in certain sectors.

Specific guidance regarding SOEs was provided in December 2006 by the National Development and Reform Commission (NDRC) when it issued a guiding opinion on state-owned assets restructuring. The opinion states that SASAC's state-owned assets should concentrate on "important industries and key areas" (i.e., strategic industries). The opinion then explained that the "important industry and key areas" shall "mainly include industries that involve national security, large and important infrastructures, important mineral resources, important public utilities and public services, and key enterprises in the pillar industries and high-tech industries."

Seven important industries and key areas were identified: defense, electric power and grid, petroleum and petrochemical, telecommunications, coal, civil aviation, and shipping. Basic and pillar industries where the state would also maintain an important role included equipment manufacturing, auto, information technology, construction, iron and steel, non-ferrous metals, chemicals, and surveying and design.

The counterpart to rapid development of key industries is maintaining low prices and ready availability of key raw materials. Not surprisingly, the cases filed by the US, the EU and Mexico against Chinese export restraints on certain raw materials involve raw materials used in some of the key industries identified in China's industrial policies – steel, aluminum and chemicals. As the USTR press release of November 28, 2009, announcing the panel request against China indicated, "The materials at issue are: bauxite, coke, fluorspar, magnesium, manganese, silicon metal, silicon carbide, yellow phosphorus and zinc, key inputs for numerous downstream products in the steel, aluminum and chemical sectors across the globe."

A corollary to keeping prices at home low is the ability to force trading partners to shutter capacity in downstream industries. For example, a study came out in December 2009, published by the European Chamber, entitled "Overcapacity in China: Causes, Impacts and Recommendations." http://www.europeanchamber.com.cn/view/static/?sid=6388. The European Chamber in China reviewed the massive problems of overcapacity in a number of important industries including steel, aluminum, cement, chemicals, refining, wind power equipment, ship building, flat glass, and photovoltaics. While the causes of overcapacity in China are varied as reviewed in the study, when coupled with export restraints on key raw materials, China can apply pressure on trading partners to make the adjustments for excess capacity created in China by limiting access at affordable prices to key raw materials or preempting development of key new technologies in the U.S. and elsewhere.

And control of key raw materials can be used to attract foreign investment by limiting access to such materials to those with a local presence. As discussed on one website, China is apparently offering to give ample supplies of rare earth minerals to companies that invest in China, even as China moves to limit or eliminate availability of product for export.

> Chinese officials have made it very clear: If foreign manufacturing companies move their facilities to China, they will be guaranteed a steady supply of rare earths. Many technology companies are reluctant to do this because they want to protect their intellectual property, but will the temptation of an endless REE supply be too much? Companies continue to move operations to China, but the tension still exists.

Clint Cox, The Anchor House, Inc. (Research on Rare Earth Elements), December 17, 2009, http://theanchorhouse.com (page 5).

When one looks at China, one sees all of the effects and/or purposes behind the wide ranging export restraints applied to rare earths and other materials. A series of articles in the last four months of 2009 reflects a range of concerns and purposes behind the draft "2009-2015 Rare Earth Industry Development Plan" from the Ministry of Industry and Information Technology:

> [A]s early as 1998, China has started to limit the export quantities of rare earth products, and implemented the differentiating principle of "forbid, encourage, and restrict:" forbid the export of rare earth raw materials; restrict oxides and metals by using export quota; encourage downstream rare earth products, such as high value-added products like magnetic materials and fluorescent powder.
>
> However, under the increasing global demand and China's increasingly reduced number of eligible export companies and export quotas, some companies with large quotas started to sell their quotas illegally. In addition, some developed countries' companies started to invest and establish factories in tungsten, antimony, and other rare earth reserves areas, bought large quantities of raw materials, processed them simply before shipping them overseas for further processing or storage, thereby effectively evaded China's export control. From 1990 to 2008, China's rare earth export grew almost 10 times, but the average export price has lowered to about 60% of the original price.

All these demonstrate that China's rare earth industry has three serious problems: overcapacity, disorderly competition, and cheap export on a large scale. It is of great urgency that we protect our rare earth resources and establish our reserve system.

MIIT's 2009-2015 Plan aims to macro-manage the rare earth industry, strengthen the control of strategic resources, and strictly control production capacity, by both administrative and market means. In the next 6 years, no new rare earth mining permit will be approved, separation of newly formed rare earth smelting companies will be strictly reviewed, and existing rare earth companies will be eliminated [by judging their performance] in the three areas of technology and equipment, environmental protection, and management.

At the same time, industry access standards will be higher, elimination of outdated capacity will be accelerated through "shut down, pause, merger, transfer." Promote merger and reorganization of companies, strengthen and enlarge rare earth industry, form leading rare earth companies, establish a "China Rare Earth OPEC," form companies with absolute dominating power in the market so that China can be the leader in controlling international market price. Of course, to accomplish this, merely depending on controlling resources and export is not enough. More importantly, we must grasp rare earth core technology patents, rare earth application market, rare earth products standards. Therefore, we must start from the technology innovation, invest more in technology, and at the same time value intellectual property, implement IP strategies, and seize the commanding ground of technology. Break out of the technology restrictions of foreign-invested enterprises, and establish our own rare earth "high-way" industry chain.

"2009-2015 Rare Earth Industry Development Plan" Has Been Passed, Hardware Business website (an electronic business website formed by Wenzhou Shengqi Internet Technology, Co., Ltd.), November 6, 2009 (unofficial translation).

China is attempting to make significant cuts to both rare earth exports. China has implemented its program to limit foreign availability of rare earths through a combination of both export duties and export quotas. These actions will raise prices outside of China by curtailing supplies and by raising import prices (with all relevant taxes or duties).

The quota for rare earth materials was 31,300 MTs in 2009, down 8.33% from 2008. This is the 5th year since China started decreasing its rare earth export.

As a corresponding policy to the annual 35,000 MTs quota from 2009 to 2015 proposed by MIIT, China will restrict its mineral annual production to

130,000 to 170,000 MTs and its rare earth smelting products' production to 120,000 to 150,000 MTs from 2009 to 2015.

"2009-2015 Rare Earth Industry Development Plan," China Suppliers' website (under the guidance of State Council Information Office, Internet Promotion Division; MOFCOM Department of Market Operation Regulation; National Development and Reform Commission, International Cooperation Center), September 4, 2009 (unofficial translation).

For 2010, the Chinese export duty and quota programs are reviewed in a series of documents issued in late 2009.

Included as Exhibit 1 to this testimony is an unofficial translation of the export duty rate chart for 2010, which is an attachment to a notice titled "State Council Customs Tariff Commission's Notice on the Implementation of the 2010 Tariff Schedule," Customs Tariff Commission Pub. [2009] No. 28, December 8, 2009. The exhibit shows export duties being assessed on 329 products in 2010 including many rare earth items (e.g., items 47, 89-92, 122-139 (export duties of $10 - 25\%$)).

Exhibit 2 to this testimony is an unofficial translation to the Ministry of Commerce of the People's Republic of China (MOFCOM) Trade Letter [2009] No. 147, December 29, 2009, "2010 1st Batch Export Quota Distribution for Rare-Earth Materials in General Trade." The total quota for the "1st batch" is 16,304 MT, with allocations given to twenty-two companies. A month and a half earlier, MOFCOM had published "Notice on Application Criteria and Procedures for 2010 Rare-Earth Materials Export Quota." MOFCOM Pub. [2009] No. 94, November 6, 2009. An unofficial translation is included as Exhibit 3.

The U.S.-China Economic and Security Review Commission (USCC) has done great work summarizing the general problem of export restrictions found in China as well as the USCC's understanding of how this problem plays out with rare earths. I have quoted the USCC at length because nobody has synthesized this data better. A complete excerpt of the USCC's views from its 2009 Report to Congress is available in Exhibit 4.

Regarding China's general export restrictions, the USCC states in its 2009 Report to Congress,

> Export restrictions or export quotas, especially on energy and raw materials, have two general effects: First, they suppress prices in the domestic market for these goods, which lowers production costs for industries that use the export-restricted materials; and second, these restrictions increase the world

price for the raw materials that are affected by limiting the world supply, thereby raising production costs in competing countries.

U.S.-China Economic and Security Review Commission (USCC) 2009 Report to Congress, at 62. Available at http://www.uscc.gov/annual report/2009/annual report full 09.pdf. While specifically addressing China's restrictions on the export of rare earth minerals, the USCC notes,

> China appears to be tightening its control over the supply of rare earth elements, valuable minerals that are used prominently in the production of such high-technology goods as flat panel screens and cell phones, and crucial green technologies such as hybrid car batteries and the special magnets used in wind turbines. USCC 2009 Report to Congress, at 63.

This reduction in supply by China is problematic because, according to the USCC, "China accounts for the vast majority—93 percent—of the world's production of rare earth minerals, and for the last three years it has been reducing the amount that can be exported." 2009 Report to Congress, at 63. China admits that rare earth elements are "the most important resource for Inner Mongolia," which contains 75 percent of China's deposits. *Id.* Accordingly, the USCC cautions that by limiting exports and controlling production, the Chinese government is attempting to "consolidate its rare earths industry, with the aim of creating a consortium of miners and processors in Inner Mongolia." *Id.* And according to the USCC, these tighter limits on exports of rare earths will place foreign manufacturers at a disadvantage compared to the domestic producers, whose access will not be so restricted. *Id.*

POLICIES AND SOLUTIONS FOR GOVERNMENT AND PRIVATE SECTOR CONSIDERATION

First and foremost, the U.S. and its trading partners should be considering a second trade action against China on the range of export restraints being imposed on rare earths (and possibly other products). The U.S. and others were concerned about China's use of export restrictions during China's negotiations for accession to the World Trade Organization. China agreed to limit the use of export taxes to 84 product categories (none of which included rare earth items) at rates no higher than included in Annex 6 of the Protocol of Accession. The fact that in 2010 China has imposed export taxes on 329 product categories, including twenty-three rare earth categories, creates a strong case of violation by China on the export taxes alone. Other violations from the use of export quotas are likely as well. Hopefully, Congressional interest will help move the Administration towards a second case on an expedited basis.

On the domestic front, it is my understanding that both the government and the private sector are taking actions to understand the nature of the potential problems as well as looking for alternative sources of supply.

For example, it is my understanding that the National Defense Authorization Act for Fiscal Year 2010, Publ. Law 111-84, requires the Comptroller General to deliver a report to the House and Senate Committees on Armed Services by April 1st this year on "rare earth materials in the defense supply chain." Section 843, 50 USC app. 2093 note. The nature of the report suggests that it is likely to provide important options that should be considered by the Government to safeguard the military needs of the country moving forward in this area. Section 843(b) is reprinted below:

(b) Matters Addressed.—The report required by subsection (a) shall address at a minimum, the following:
    (1) An analysis of the current and projected domestic and worldwide availability of rare earths for use in defense systems, including an analysis of projected availability of these materials in the export market.
    (2) An analysis of actions or events outside the control of the Government of the United States that could restrict the access of the Department of Defense to rare earth materials, such as past procurements and attempted procurements of rare earth mines and mineral rights.
    (3) A determination as to which defense systems are currently dependent on, or projected to become dependent on, rare earth materials, particularly neodymium iron boron magnets, whose supply could be restricted—
        (A) by actions or events identified pursuant to paragraph (2); or
        (B) by other actions or events outside the control of the Government of the United States.
    (4) The risk to national security, if any, of the dependencies (current or projected) identified pursuant to paragraph (3).
    (5) Any steps that the Department of Defense has taken or is planning to take to address any such risk to national security.

(6) Such recommendations for further action to address the matters covered by the report as the Comptroller General considers appropriate.

Historically, the U.S. maintained a strategic stockpile of critical materials for national defense. Presumably, one of the issues that will be addressed in the report is the extent to which stockpiling rare earth materials is appropriate or feasible.

The Chairman and Members of this Subcommittee might want to advocate creation of a similar report for the civilian sector. Such a report would obviously be helpful to Members of Congress in understanding the challenges facing the American economy from the current reliance on China as the source of supply and what legislative approaches might be pursued to safeguard our commercial and military interests.

Press accounts suggest that in recent years there has been renewed interest in developing rare earth mineral resources outside of China and that several mines are in the process of being reactivated or developed. See, e.g., "New USGS Rare Earth Report Includes Thorium Energy, Inc.," Earth Times, Oct. 8, 2009; http://www.earthtimes.org/articles/show/new-usgs-rare-earth-report-includes-thoriumenergy-inc,99 1131. shtml; "Canadian firms set up search for rare-earth metals," New York Times, Sept. 9, 2009, http://www.nytimes.com/2009/09/10/business/global/10mineral.html?r=1&scp=10&sq=br

Possible American sources of rare-earths include a separation plant at Mount Pass, CA. Bastnasite concentrates and other rare-earth intermediates and refined products continue to be sold from mine stocks at Mountain Pass. Exploration for rare earths continued in 2009; however, global economic conditions were not as favorable as in early 2008. Economic assessments continued at Bear Lodge in Wyoming; Diamond Creek in Idaho; Elk Creek in Nebraska; and Lemhi Pass in Idaho-Montana.

Thus, government and the private sector may have additional sources of supply of rare earths beyond China, although the challenge may be overall cost of supply, particularly in countries like the U.S. or Canada where environmental needs are more likely to be addressed at present than in China.

Presumably, the government, under CFIUS, can help ensure that mines in the U.S. are not purchased by foreign interests whose governments have limited supply to U.S. users and that new mines receive priority attention in terms of various government licenses and reviews.

I note that the Senate Committee on Energy and Natural Resources held a hearing last summer on mining law reform. The hearing had a number of witnesses who talked about the ability to improve the U.S. ability to supply more of its rare earth mineral needs and what challenges they faced based on various pending bills. Mining Law Reform, S. Hrg. 111-116, 111th Cong., 1st Sess. (July 14, 2009)(S.796; S.140). Certainly, the Congress will want to be sure that any legislation balances our needs for access to critical raw materials with the other concerns prompting legislative modifications.

Finally, the USGS indicates that for most rare earth minerals there are substitute products available, although known substitutes are less effective than the rare earth minerals. The U.S. government can support research efforts into the development of alternative solutions to current rare earth needs both directly through basic and applied research and through tax policies and other actions to support private sector research.

Thank you for the opportunity to appear today. I would be pleased to respond to any questions.

Exhibit 1. 出口商品税率表 Export Duty Rate Chart (2010)

序号 No.	EX ①	税则号列 Tariff No.	商品名称（简称） Product Name	出口税率 (%) Export Duty Rate (%)	2010年暂定税率(%) 2010 Interim Export Duty Rate (%)	2010年特别出口税率 (%) 2010 Special Export Duty Rate (%)
1		03019210	鳗鱼苗 Eels fry	20	10	
2		05061000	经酸处理的骨胶原及骨 Ossein and bones treated with acid	40		
3		05069011	含牛羊成分的骨粉及骨废料 Powder and waste of bones of bovine and sheep	40		
4		05069019	其他骨粉及骨废料 Other powder and waste of bones	40		
5	ex	05069090	其他骨及角柱(已脱胶骨，角柱除外) Other bones and horn-cores (other than degelatinized bones and horn-cores)	40		
6	ex	05069090	已脱胶骨，角柱 Degelatinized bones and horn-cores	40	0	
7		25041010	鳞片状天然石墨 Natural graphite in flakes		20	
8		25041099	其它粉末状天然石墨 Other natural graphite in power		20	
9		25049000	其他天然石墨 Other natural graphite		20	

Exhibit 1. (Continued)

序号 No.	EX ①	税则号列 Tariff No.	商品名称（简称）Product Name	出口税率（%）Export Duty Rate (%)	2010年暂定税率(%) 2010 Interim Export Duty Rate (%)	2010年特别出口税率（%）2010 Special Export Duty Rate (%)
10		25085000	红柱石,蓝晶石及硅线石，不论是否煅烧 Andalusite, kyanite, and sillimanite, whether calcined or not		10	
11		25086000	富铝红柱石 Mullite		10	
12		25101010	未碾磨磷灰石 Ungrounded apatite		35	
13		25101090	未碾磨天然磷酸钙、天然磷酸铝钙及磷酸盐白垩，磷灰石除外 Ungrounded natural calcium phosphates, natural aluminum calcium phosphates, and phosphatic chalk other than apatite		35	
14		25102010	已碾磨磷灰石 Ground apatite		35	
15		25102090	已碾磨天然磷酸钙、天然磷酸铝钙及磷酸盐白垩，磷灰石除外 Ground natural calcium phosphates, natural aluminum calcium phosphates, and phosphatic chalk other than apatite		35	
16		25111000	天然硫酸钡(重晶石) Natural barium sulphate (barites)		10	

Exhibit 1. (Continued)

序号 No.	EX ①	税则号列 Tariff No.	商品名称（简称） Product Name	出口税率（%）Export Duty Rate (%)	2010年暂定税率(%) 2010 Interim Export Duty Rate (%)	2010年特别出口税率（%）2010 Special Export Duty Rate (%)
17		25112000	天然碳酸钡(毒重石)，不论是否煅烧 Natural barium carbonate(witherite),whether calcined or not		10	
18		25191000	天然碳酸镁（菱镁矿）Natural magnesium carbonate (magnesite)		5	
19		25199010	熔凝镁氧矿 Fused magnesia		10	
20		25199020	烧结镁氧矿（重烧镁）Dead-burned (sintered) magnesia		10	
21		25199030	碱烧镁（轻烧镁）Light-burned magnesia		5	
22		25199099	非纯氧化镁 Other magnesium oxide, other than chemically pure magnesium oxide		5	
23		25261020	未破碎及未研粉的滑石,不论是否粗加修整或切割成矩形板块 Talc, not crushed, not powdered, whether or not roughly trimmed or merely cut into blocks or slabs or a rectangular shape		10	

Exhibit 1. (Continued)

序号 No.	EX ①	税则号列 Tariff No.	商品名称(简称) Product Name	出口税率(%) Export Duty Rate (%)	2010年暂定税率(%) 2010 Interim Export Duty Rate (%)	2010年特别出口税率(%) 2010 Special Export Duty Rate (%)
24	ex	25262020	已破碎或已研粉的天然滑石(体积百分比90%及以上的产品粒度小于等于1 8微米的滑石粉除外)Crushed or powdered talc (other than that containing talc 90% or more by volume and of a granularity of 18mm or less)		10	
25	ex	25262020	体积百分比90%及以上的产品粒度小于等于1 8微米的滑石粉 Crushed or powdered talc, containing talc 90% or more by volume and of a granularity of 18mm or less		5	
26		25292100	按重量计氟化钙含量≤97%的萤石 Fluorspar, containing by weight 97% or less of calcium fluoride		15	
27		25292200	按重量计氟化钙含量 >97%的萤石 Fluorspar, containing by weight more than 97% of calcium fluoride		15	
28		25309020	稀土金属矿 Ores of rare earth metals		15	
29	ex	25309099	其他氧化镁含量在70% (含70%)以上的矿产品 Other mineral products with 70% or more of magnesia		5	

Exhibit 1. (Continued)

序号 No.	EX ①	税则号列 Tariff No.	商品名称（简称）Product Name	出口税率（%）Export Duty Rate (%)	2010年暂定税率(%) 2010 Interim Export Duty Rate (%)	2010年特别出口税率（%）2010 Special Export Duty Rate (%)
30		25301010	未膨胀的绿泥石 Chlorites, unexpanded		10	
31		26011110	平均粒径小于0.8毫米的未煅烧铁矿*砂及其精矿*；但焙烧黄铁矿除外 Iron ores and concentrates, other than roasted iron pyrites, non-agglomerated, of a granularity of less than 0.8mm		10	
32		26011120	平均粒径不小于0.8毫米，但不大于6.3毫米的未煅烧铁矿*砂及其精矿*；但焙烧黄铁矿除外 Iron ores and concentrates, other than roasted iron pyrites, non-agglomerated, of a granularity of 0.8mm or more, but not exceeding 6.3mm		10	
33		26011190	平均粒径大于6.3毫米的未烧结铁矿*砂及其精矿*，但焙烧黄铁矿除外 Iron ores and concentrates, other than roasted iron pyrites, non-agglomerated, of a granularity of more than 6.3mm		10	

Exhibit 1. (Continued)

序号 No.	EX ①	税则号列 Tariff No.	商品名称（简称）Product Name	出口税率(%) Export Duty Rate (%)	2010年暂定税率(%) 2010 Interim Export Duty Rate (%)	2010年特别出口税率（%）2010 Special Export Duty Rate (%)
34		26011200	已烧结铁矿砂及其精矿 Agglomerated iron ores and concentrates		10	
35		26012000	焙烧黄铁矿 Roasted iron pyrites		10	
36		26020000	锰矿砂及其精矿，包括以干重计含锰量在20%及以上的锰铁砂及其精矿 Manganese ores and concentrates, including ferruginous manganese ores and concentrates with a manganese content of 20% or more, calculated on the dry weight		15	
37		26030000	铜矿砂及其精矿 Copper ores and concentrates		10	
38		26040000	镍矿砂及其精矿 Nickel ores and concentrates		15	
39		26050000	钴矿砂及其精矿 Cobalt ores and concentrates		15	
40		26070000	铅矿砂及其精矿 Lead ores and concentrates	30		
41	ex	26080000	锌矿砂及其精矿（氧化锌含量大于80%的灰色饲料氧化锌除外）Zinc ores and concentrates (other than grey feed grade zinc oxide, containing more than 80% of zinc oxide by weight)	30		

Exhibit 1. (Continued)

序号 No.	EX ①	税则号列 Tariff No.	商品名称（简称）Product Name	出口税率(%) Export Duty Rate (%)	2010年暂定税率(%) 2010 Interim Export Duty Rate (%)	2010年特别出口税率(%) 2010 Special Export Duty Rate (%)
42	ex	26080000	灰色饲料氧化锌（氧化锌ZnO含量大于80%）Grey feed grade zinc oxide, containing more than 80% of zinc oxide by weight	30	0	
43		26090000	锡矿砂及其精矿 Tin ores and concentrates	50	20	
44		26100000	铬矿砂及其精矿 Chromium ores and concentrates		15	
45		26110000	钨矿砂及其精矿 Tungsten ores and concentrates	20		
46		26121000	铀矿砂及其精矿 Uranium ores and concentrates		10	
47		26122000	钍矿砂及其精矿 Thorium ores and concentrates		10	
48		26131000	已焙烧钼矿砂及其精矿 Molybdenum ores and concentrates, roasted		15	
49		26139000	其他钼矿砂及其精矿 Molybdenum ores and concentrates, other		15	
50		26140000	钛矿砂及其精矿 Titanium ores and concentrates		10	

Exhibit 1. (Continued)

序号 No.	EX号 ①	税则号列 Tariff No.	商品名称（简称）Product Name	出口税率（%）Export Duty Rate (%)	2010年暂定税率(%) 2010 Interim Export Duty Rate (%)	2010年特别出口税率（%）2010 Special Export Duty Rate (%)
51		26151000	锆矿砂及其精矿 Zirconium ores and concentrates		10	
52		26159010	水合钽铌原料（钽铌富集物）Hydrated tantalum/niobium materials or enriched materials from tantalum/niobium ore			
53		26159090	其他钽铌矿砂及其精矿 tantalum/niobium ores and concentrates, other			
54		26161000	银矿砂及其精矿 Silver ores and concentrates			
55		26169000	其他贵金属矿砂及其精矿 Precious metal ores and concentrates, other			
56		26171010	生锑（锑精矿，选矿产品）Crude antimony (antimony concentrates which are mineral products)			
57		26171090	其他锑矿砂及其精矿 Antimony ores and concentrates, other			
58		26179010	朱砂(辰砂) Cinnabar			
59		26179090	其他矿砂及其精矿 Other ores and concentrates, other			

Exhibit 1. (Continued)

序号 No.	EX ①	税则号列 Tariff No.	商品名称（简称） Product Name	出口税率(%) Export Duty Rate (%)	2010年暂定税率(%) 2010 Interim Export Duty Rate (%)	2010年特别出口税率（%) 2010 Special Export Duty Rate (%)
60		26180010	冶炼钢铁产生的锰渣 Granulated slag (slag sand) from the manufacture of iron or steel, containing mainly manganese			
61		26180090	冶炼钢铁产生的其他粒状熔渣(熔渣砂) Granulated slag (slag sand) from the manufacture of iron or steel, other			
62		26190000	冶炼钢铁产生的熔渣、浮渣、氧化皮及其他废料 Slag, dross (other than granulated slag),scalings and other waste from the manufacture of iron or steel			
63		26201100	含硬锌的矿灰及残渣 Slag, ash, and residues, containing mainly hard zinc spelter			
64		26201900	含其他锌的矿灰及残渣 slag, ash, and residues,containing mainly other zinc			
65		26202100	含铅汽油的淤渣及含铅抗震化合物的淤渣 Leaded gasoline sludges and leaded anti-knock compound sludges			
66		26202900	其他主要含铅的矿灰及残渣 Other zinc-containing ore slag, ore ash and residues, containing mainly lead,other			

Exhibit 1. (Continued)

序号 No.	EX ①	税则号列 Tariff No.	商品名称（简称）Product Name	出口税率(%) Export Duty Rate (%)	2010年暂定税率(%) 2010 Interim Export Duty Rate (%)	2010年特别出口税率（%）2010 Special Export Duty Rate (%)
67		26203000	主要含铜的矿灰及残渣 Other zinc-containing ore slag, ore ash and residues, containing mainly copper			
68		26206000	含砷、汞、铊及其混合物，用于提取或生产砷、汞、铊及其化合物的矿灰及残渣 Other zinc-containing ore slag, ore ash and residues, containing arsenic, mercury, thallium or their mixtures, of a kind used for the extraction of arsenic, mercury, thallium, or for the manufacture of their chemical compounds			
69		26209100	含锑、铍、镉、铬及其混合合物的矿灰及残渣 Other zinc-containing ore slag, ore ash and residues, containing antimony, beryllium, cadmium, chromium, or their mixtures		10	
70		26209910	主要含钨的矿灰及残渣 Other zinc-containing ore slag, ore ash and residues, containing mainly tungsten		10	
71		26209990	含其他金属及化合物的矿灰及残渣 Other zinc-containing ore slag, ore ash and residues, other		10	

Exhibit 1. (Continued)

序号 No.	EX ①	税则号列 Tariff No.	商品名称（简称）Product Name	出口税率(%) Export Duty Rate (%)	2010年暂定税率(%) 2010 Interim Export Duty Rate (%)	2010年特别出口税率（%）2010 Special Export Duty Rate (%)
72		27011100	未制成型的无烟煤，不论是否粉化 Coal, whether or not pulverized, but not agglomerated,anthracite		10	
73		27011210	未制成型的炼焦烟煤，不论是否粉化 Coal, whether or not pulverized, but not agglomerated, bituminous coal, coking coal		10	
74		27011290	未制成型的其他烟煤，不论是否粉化 Coal, whether or not pulverized, but not agglomerated, bituminous coal, other		10	
75		27011900	未制成型的其他煤，不论是否粉化 Coal, whether or not pulverized, but not agglomerated,other coal		10	
76		27012000	煤砖、煤球及类似用煤制固体燃料 Briquettes, ovoids, and similar solid fuels manufactured from coal		10	
77		27021000	褐煤 Lignite		10	
78		27022000	制成型的褐煤 Agglomerated lignite		10	
79		27030000	泥煤（包括肥料用泥煤)不论是否成型Peat (including peat litter) whether or not agglomerated		10	

Exhibit 1. (Continued)

序号 No.	EX ①	税则号列 Tariff No.	商品名称（简称）Product Name	出口税率（%）Export Duty Rate (%)	2010年暂定税率(%) 2010 Interim Export Duty Rate (%)	2010年特别出口税率（%）2010 Special Export Duty Rate (%)
80		27040010	煤制焦炭及半焦炭不论是否成型 Coke and semi-coke of coal, whether or not agglomerated		40	
81	ex	27060000	从煤、褐煤、或泥煤蒸馏所得的焦油及矿物焦油， 不论是否脱水或部分蒸馏，包括再造焦油（含蒽油≥50%及沥青≥40%的"炭黑油"除外）Tar distilled from coal, from lignite or from peat, and other mineral tars, whether or not dehydrated or partially distilled, including reconstituted tars (other than anthracene, containing oil ≥ 50% and asphalt ≥ 40% "black oil")			
82		27071000	粗苯 Benzole		10	
83		27090000	石油原油及从沥青矿物提取的原油 Petroleum oils and oils obtained from bituminous minerals, crude		5	
84		28046900	按重量计硅含量小于99.99%的硅 Silicon, containing by weight less than 99.99% of silicon		15	

Exhibit 1. (Continued)

序号 No.	EX ①	税则号列 Tariff No.	商品名称（简称）Product Name	出口税率(%) Export Duty Rate (%)	2010年暂定税率(%) 2010 Interim Export Duty Rate (%)	2010年特别出口税率(%) 2010 Special Export Duty Rate (%)
85		28047010	黄磷(白磷) Yellow phosphorus (white phosphorus)	20		
86		28047090	其他磷 Other phosphorus	20	10	
87		28053011	钕 Neodymium		15	
88		28053012	镝 Dysprosium		25	
89		28053013	铽 Terbium		25	
90		28053019	其他未相互混合或熔合的稀土金属、钪及钇 Rare-earth materials, scandium and yttrium, not intermixed or interalloyed, other		25	
91		28053021	已相互混合或熔合的稀土金属、钪及钇，电池级 Rare-earth materials, scandium and yttrium, intermixed or interalloyed, battery grade		25	
92		28053029	其他已相互混合或熔合的稀土金属、钪及钇 Rare-earth materials, scandium and yttrium, intermixed or interalloyed, other		25	
93		28092019	磷酸 Phosphoric acid		7	

Exhibit 1. (Continued)

序号 No.	EX ①	税则号列 Tariff No.	商品名称（简称）Product Name	出口税率(%) Export Duty Rate (%)	2010年暂定税率(%) 2010 Interim Export Duty Rate (%)	2010年特别出口税率（%）2010 Special Export Duty Rate (%)
94		28111100	氢氟酸（氟化氢）Hydrofluoric acid (hydrogen fluoride)		15	
95		28141000	氨 Anhydrous ammonia		7	
96		28142000	氨水 Ammonia in aqueous solution		7	
97		28220090	其他钴的氧化物及氢氧化物；商品氧化钴 Cobalt oxides and hydroxides; commercial cobalt oxides, other		10	
98		28253010	五氧化二钒 Divanadium pentaoxide		5	
99	ex	28256000	锗的氧化物 Germanium oxides		5	
100		28257000	钼的氧化物及氢氧化物 Molybdenum oxides and hydroxides		5	
101		28259011	钨酸 Tungstic acid		5	
102		28259012	三氧化钨 Tungstic oxide		5	
103		28259019	其他钨的氧化物和氢氧化物 Tungstic oxides and hydroxides, other		5	
104		28261290	其他氟化铝 Aluminum fluoride, other		5	

Exhibit 1. (Continued)

序号 No.	EX ①	税则号列 Tariff No.	商品名称（简称）Product Name	出口税率(%) Export Duty Rate (%)	2010年暂定税率(%) 2010 Interim Export Duty Rate (%)	2010年特别出口税率（%）2010 Special Export Duty Rate (%)
105		28261910	铵的氟化物 Fluorides, of ammonium		5	
106		28261920	钠的氟化物 Fluorides, of sodium		5	
107		28261990	其他氟化物 Fluorides, other		5	
108	ex	28269090	氟钽酸钾 Potassium tantalifluoride	30		
109		28331100	硫酸钠 Disodium sulphate		5	
110		28342110	肥料用硝酸钾 Potassium nitrate, for use as fertilizer		7	
111		28417010	钼酸铵 Ammonium molybdates		5	
112		28417090	其他钼酸盐 Molybdates, other		5	
113		28418010	仲钨酸铵 Ammonium paratungstate		5	
114		28418020	钨酸钠 Sodium tungstate		5	
115		28418030	钨酸钙 Calcium wolframate		5	
116		28418040	偏钨酸铵 Ammonium metatungstate		5	
117		28418090	其他钨酸盐 Tungstates (wolframates), other		5	

Exhibit 1. (Continued)

序号 No.	EX ①	税则号列 Tariff No.	商品名称（简称）Product Name	出口税率(%) Export Duty Rate (%)	2010年暂定税率(%) 2010 Interim Export Duty Rate (%)	2010年特别出口税率(%) 2010 Special Export Duty Rate (%)
118		28461010	氧化铈 Cerium oxide		15	
119		28461020	氢氧化铈 Cerium hydroxide		15	
120		28461030	碳酸铈 Cerium carbonate		15	
121		28461090	铈的其他化合物 Ceric compounds, other		15	
122		28469011	氧化钇 Yttrium oxide		25	
123		28469012	氧化镧 Lanthanum oxide		15	
124		28469013	氧化钕 Neodymium oxide		15	
125		28469014	氧化铕 Europium oxide		25	
126		28469015	氧化镝 Dysprosium oxide		25	
127		28469016	氧化铽 terbium oxide		25	
128	ex	28469019	其他氧化稀土（灯用红粉除外）Other rare-earth oxide (other than red powder for lumination)		15	
129		28469021	氯化铽 Terbium chlorinates		25	
130		28469022	氯化镝 Dysprosium chlorinates		25	

Exhibit 1. (Continued)

序号 No.	EX ①	税则号列 Tariff No.	商品名称（简称）Product Name	出口税率(%) Export Duty Rate (%)	2010年暂定税率(%) 2010 Interim Export Duty Rate (%)	2010年特别出口税率（%）2010 Special Export Duty Rate (%)
131		28469028	混合氯化稀土 Mixture of rare-earth chlorides		15	
132		28469029	未混合氯化稀土 Unmixed rare-earth chlorides		15	
133		28469030	氟化稀土 Rare-earth fluorides		15	
134		28469041	碳酸镧 Lanthanum carbonates		15	
135		28469042	碳酸铽 Terbium carbonates		25	
136		28469043	碳酸镝 Dysprosium carbonates		25	
137		28469048	混合碳酸稀土 Mixture of rare-earth carbonates		15	
138		28469049	未混合碳酸稀土 Unmixed rare-earth carbonates		15	
139		28469090	稀土金属，钇，钪的其他化合物 Compounds of rare-earth metals, of yttrium or of scandium, other		25	
140		28499020	碳化钨 Tungsten carbide		5	
141		29022000	苯 Benzene	40	0	

Exhibit 1. (Continued)

序号 No.	EX ①	税则号列 Tariff No.	商品名称（简称）Product Name	出口税率（%）Export Duty Rate (%)	2010年暂定税率(%) 2010 Interim Export Duty Rate (%)	2010年特别出口税率（%）2010 Special Duty Rate (%)
142		31021000	尿素② Urea		旺季(2-6月,9月16日-10月15日)：35%；淡季(1月,7月1日-9月15日,10月16日-12月31日)：7%。当出口价格不高于基准价格时，7%。当出口价格高于基准价格时，税率=(1.07-基准价格/出口价格)*100%(基准价格按2.3元/公斤计算) During peak season (February to June, September 16 to October 15): 35%. During off sea-son (January, July 1 to Sept-ember 15, October 16 to December 31): when export price does not exceed base price, 7%; when export price exceeds base price, duty rate = (1.07-base price/export price)* 100% (base price is RMB2. 30/kg)	旺季：75% During peak season: 75%
143		31024000	硝酸铵与碳酸钙或其他无肥效无机物的混合物 Mixtures of ammonium nitrate with calcium carbonate or other inorganic nonfertilizing substances			
144		31025000	硝酸钠 Sodium nitrate		7	
145		31026000	硝酸钙和硝酸铵的复盐及混合物 Double salts and mixtures of calcium nitrate and ammonium nitrate		7	

Exhibit 1. (Continued)

序号 No.	EX ①	税则号列 Tariff No.	商品名称（简称）Product Name	出口税率（%）Export Duty Rate (%)	2010年暂定税率(%) 2010 Interim Export Duty Rate (%)	2010年特别出口税率（%）2010 Special Export Duty Rate (%)
146		31028000	尿素及硝酸铵混合物的水溶液或氨水溶液 Mixtures of urea and ammonium nitrate in aqueous or ammoniacal solution		7	
147		31029010	氰氨化钙 Calcium cyanamide		7	
148		31029090	其他矿物氮肥及化学氮肥，包括上述子目未列名的混合物 Mineral or chemical fertilizers, nitrogenous, other, including mixtures not specified in the foregoing subheadings		7	
149		31031010	重过磷酸钙 Triple superphosphates		7	
150		31031090	其他过磷酸钙 Other superphosphates		7	
151		31039000	其他矿物磷肥或化学磷肥 Mineral or chemical fertilizers, phosphatic, other		7	
152		31042090	其他氯化钾 Mineral or chemical fertilizers, potassic, other		30	75
153		31043000	硫酸钾 Potassium sulphate		30	75
154		31049010	光卤石、钾盐及其他天然粗钾盐 Carnallite, sylvite, and other crude natural potassiumsalts		30	75

Exhibit 1. (Continued)

序号 No.	EX ①	税则号列 Tariff No.	商品名称（简称）Product Name	出口税率（%）Export Duty Rate (%)	2010年暂定税率(%) 2010 Interim Export Duty Rate (%)	2010年特别出口税率（%）2010 Special Export Duty Rate (%)
155		31049090	其他矿物钾肥及化学钾肥 Mineral or chemical fertilizers, potassic,other		30	75
156		31051000	制成片状及类似形状或每包毛重不超过10公斤的31章各货品 Goods of Chapter 31 (Fertilizers) in tablets or similar forms or in packages of a gross weight not exceeding 10kg		7	
157		31052000	三元复合肥 Mineral or chemical fertilizers containing the three fertilizing elements nitrogen, phosphorus, and potassium		1-9月:35%; 10-12月:20% January to September: 35%. October to December:20%	75
158		31053000	磷酸氢二铵② Diammonium hydrogenorthophosphate (diammonium phosphate)		旺季(2-5月,9月1日-10月15日)：35%；淡季(1月,6-8月,10月16日-12月31日)：当出口价格不高于基准价格时，7%；当出口价格高于基准价格时，税率=(1.07-基准价出口价格)*100%（基准价格按4.0元/公斤计算）During peak season (February to May, September 1 to October 15)：35%. During off season (January, June to August, October 16 to December 31): when export price does not exceed base price, 7%; when export price exceeds base price,duty rate = (1.07-base price/export price)*100% (base price is RMB4.00/kg)	旺季：7 5% During peak season: 75%.

Exhibit 1. (Continued)

序号 No.	EX ①	税则号列 Tariff No.	商品名称（简称）Product Name	出口税率(%) Export Duty Rate (%)	2010年暂定税率(%) 2010 Interim Export Duty Rate (%)	2010年特别出口税率（%）2010 Special Export Duty Rate (%)
159		31054000	磷酸二氢铵及磷酸二氢铵与磷酸氢二铵的混合物②Ammonium dihydrogenorthophosphate (monoammonium phosphate) and mixtures thereof with diammonium hydrogenorthophosphate (diammonium phosphate)		旺季(2-5月,9月1日-10月15日)：35%；淡季(1月，6-8月,10月16-12月31日)：当出口价格不75	旺季：7 5% During peak season:
160		31055100	含有硝酸盐及磷酸盐的肥料 Other mineral or chemical fertilizers containing nitrates and phosphates		%. 高于基准价格时，7 %；当出口价格高于基准价格时，税率=(1.07-基准价格/出口价格）* 100%（基准价格按3.7元/公斤计算）During peak season (February to May, September 1 to October 15):35%. During off season (January, June to August, October 16 to December 31): when export price does not exceed base price, 7%; when export price exceeds base price,duty rate = (1.07-base price/export price)*100% (base price is RMB3.70/kg)	7

Exhibit 1. (Continued)

序号 No.	EX ①	税则号列 Tariff No.	商品名称（简称） Product Name	出口税率（%）Export Duty Rate (%)	2010年暂定税率(%) 2010 Interim Export Duty Rate (%)	2010年特别出口税率（%）2010 Special Duty Rate (%)
161		31055900	其他含氮磷两种肥效元素的矿物肥料或化学肥料 Other mineral or chemical fertilizers containing the two fertilizing elements nitrogen and phosphorus		7	
162		31056000	含磷钾两种元素的肥料 Mineral or chemical fertilizers containing the two fertilizing elements phosphorus and potassium		7	
163		31059000	其他肥料 Other fertilizers		7	
164		38249091	含滑石50%以上的混合物 Compounds containing more than 50% of talc by weight		10	
165		41039011	经退鞣处理的山羊板皮 Dried hides and skins of goats, have undergoing a reversible tanning process	20		
166		41039019	山羊板皮，经退鞣处理的除外 Dried hides and skins of goats, other	20		
167		44012100	针叶木木片或木粒 Wood in chips or particles, coniferous		15	

Exhibit 1. (Continued)

序号 No.	EX ①	税则号列 Tariff No.	商品名称（简称）Product Name	出口税率（%）Export Duty Rate (%)	2010年暂定税率(%) 2010 Interim Export Duty Rate (%)	2010年特别出口税率（%）2010 Special Export Duty Rate (%)
168		44012200	非针叶木木片或木粒 Wood in chips or particles, non-coniferous	15		
169		44091010	针叶木地板条（块）Coniferous floor board strips		10	
170		44092910	其他非针叶木地板条 Other non-coniferous floor board strips		10	
171		44190031	木制一次性筷子 One-time chopsticks of wood		10	
172		44219021	木制圆签、圆棒、冰果棒、压舌片及类似一次性制品 Round toothpick, round stick, ice fruit stick, tongue-pressing plate, and similar one-time products, of wood		10	
173		47010000	机械木浆 Mechanical wood pulp		10	
174		47020000	化学木浆，溶解级 chemical wood pulp, dissolving grades		10	
175		47031100	未漂白针叶木碱木浆或硫酸盐木浆 Chemical wood pulp, soda or sulphate, other than dissolving grades, unbleached, coniferous		10	

Exhibit 1. (Continued)

序号 No.	EX ①	税则号列 Tariff No.	商品名称（简称） Product Name	出口税率（%） Export Duty Rate (%)	2010年暂定税率(%) 2010 Interim Export Duty Rate (%)	2010年特别出口税率（%） 2010 Special Export Duty Rate (%)
176		47031900	未漂白非针叶木碱木浆或硫酸盐木浆 Chemical wood pulp,soda or sulphate,other than dissolving grades, unbleached, non-coniferous		10	
177		47032100	漂白针叶木碱木浆或硫酸盐木浆 Chemical wood pulp, soda or sulphate, other than dissolving grades, bleached, coniferous		10	
178		47032900	漂白非针叶木碱木浆或硫酸盐木浆 Chemical wood pulp, soda or sulphate, other than dissolving grades, bleached, non-coniferous		10	
179		47041100	未漂白的针叶木亚硫酸盐木浆 Chemical wood pulp, sulphite, other than dissolving grades, unbleached, coniferous		10	
180		47041900	未漂白的非针叶木亚硫酸盐木浆 Chemical wood pulp, sulphite, other than dissolving grades, unbleached,non-coniferous		10	
181		47042100	漂白的针叶木亚硫酸盐木浆 Chemical wood pulp, sulphite, other than dissolving grades, bleached, coniferous		10	

Exhibit 1. (Continued)

序号 No.	EX ①	税则号列 Tariff No.	商品名称（简称） Product Name	出口税率（%）Export Duty Rate (%)	2010年暂定税率(%) 2010 Interim Export Duty Rate (%)	2010年特别出口税率（%）2010 Special Export Duty Rate (%)
182		47042900	漂白的非针叶树亚硫酸盐木浆 Chemical wood pulp, sulphite, other than dissolving grades, bleached, non-coniferous	10		
183		47050000	半化学木浆 Wood pulp obtained by a combination of mechanical and chemical pulping process (semi-chemical wood pulp)	10		
184		47062000	从回收纸或纸板提取的纤维浆 Pulps of fibers derived from recovered (waste and scrap) paper or paperboard	10		
185		47063000	竹浆 Pulps of bamboo	10		
186		47069100	其他纤维状纤维素机械浆 Other mechanical pulp derived from fibrous cellulosic material	10		
187		47069200	其他纤维状纤维素化学浆 Other chemical pulp derived from fibrous cellulosic material	10		
188		47069300	其他纤维状纤维素半化学浆 Other semi-chemical pulp derived from fibrous cellulosic material	10		

Exhibit 1. (Continued)

序号 No.	EX ①	税则号列 Tariff No.	商品名称（简称）Product Name	出口税率(%) Export Duty Rate (%)	2010年暂定税率(%) 2010 Interim Export Duty Rate (%)	2010年特别出口税率（%）2010 Special Export Duty Rate (%)
189		72011000	非合金生铁，含磷量小于或等于 0.5% Non-alloy pig iron containing by weight 0.5% or less of phosphorus	20	25	
190		72012000	非合金生铁，含磷量大于 0.5% Non-alloy pig iron containing by weight more than 0.5% of phosphorus	20	25	
191		72015000	合金生铁 Alloy pig iron	20	25	
192		72021100	锰铁,含碳量>2% Ferro-manganese, containing by weight more than 2% of carbon	20		
193		72021900	锰铁,含碳量≤2% Ferro-manganese, containing by weight 2% or less of carbon	20		
194		72022100	硅铁,含硅量>55% Ferro-silicon, containing by weight more than 55% of silicon	25		
195		72022900	硅铁,含硅量≤55% Ferro-silicon, containing by weight 55% or less of silicon	25		

Exhibit 1. (Continued)

序号 No.	EX ①	税则号列 Tariff No.	商品名称（简称） Product Name	出口税率(%) Export Duty Rate (%)	2010年暂定税率(%) 2010 Interim Export Duty Rate (%)	2010年特别出口税率(%) 2010 Special Export Duty Rate (%)
196		72023000	硅锰铁 Ferro-silico-manganese	20		
197		72024100	铬铁,含碳量＞4% Ferro-chromium, containing by weight more than 4% of carbon	40	20	
198		72024900	铬铁,含碳量≤4% Ferro-chromium, containing by weight 4% or less of carbon	40	20	
199		72025000	硅铬铁 Ferro-silico-chromium		20	
200		72026000	镍铁 Ferro-nickel		20	
201		72027000	钼铁 Ferro-molybdenum		20	
202		72028010	钨铁 Ferro-tungsten		20	
203		72028020	硅钨铁 Ferro-silico-tungsten		20	
204		72029100	钛铁及硅钛铁 Ferro-titanium and ferro-silico-titanium		20	
205		72029290	其他钒铁 Other ferro-vanadium		20	
206		72029300	铌铁 Ferro-niobium		20	

Exhibit 1. (Continued)

序号 No.	EX ①	税则号列 Tariff No.	商品名称（简称）Product Name	出口税率（%）Export Duty Rate (%)	2010年暂定税率(%) 2010 Interim Export Duty Rate (%)	2010年特别出口税率（%）2010 Special Export Duty Rate (%)
207		72029919	其他钕铁硼合金 Other ferro-neodymium-boron alloy		20	
208		72029990	其他铁合金 Other ferro-alloys		20	
209		72031000	直接从铁矿还原的铁产品 Ferrous products obtained by direct reduction of iron ore		25	
210		72039000	其他铁，海绵铁,产品纯度＞99.94% Other ferrous products, spongy ferrous products, iron having a purity of more than 99.94% by weight		25	
211		72041000	铸铁废碎料 Waste and scrap of cast iron	40		
212		72042100	不锈钢废碎料 Waste and scrap of stainless steel	40		
213		72042900	其他合金钢废碎料 Other waste and scrap of alloy steel	40		
214		72043000	镀锡钢铁废碎料 Waste and scrap of tinned iron or steel	40		
215		72044100	机械加工中产生的废料 Waste and scrap of iron and steel produced in mechanical processing	40		

Exhibit 1. (Continued)

序号 No.	EX ①	税则号列 Tariff No.	商品名称 (简称) Product Name	出口税率 (%) Export Duty Rate (%)	2010年暂定税率(%) 2010 Interim Export Duty Rate (%)	2010年特别出口税率 (%) 2010 Special Export Duty Rate (%)
216		72044900	其他钢铁废碎料 Other waste and scrap of iron and steel	40		
217		72045000	供再熔的碎料钢铁锭 Remelting scrap ingots	40		
218	ex	72051000	生铁、镜铁及钢铁颗粒(不带球角钢砂除外) Granules of pig iron, spiegeleisen, iron or steel (other than angular steel grit containing more than 80% of angular granules without spherical arc) 颗粒数量大于80%的棱角钢砂除外		25	
219		72052900	生铁、镜铁及其他钢铁粉末 Powders of pig iron, spiegeleisen, iron or steel		25	
220		72061000	铁及非合金钢锭 Iron and non-alloy steel in ingots		25	
221		72069000	其他初级形状的铁及非合金钢 Iron and non-alloy steel in other primary forms		25	
222		72071100	宽度＜厚度两倍的矩形截面钢坯,C<0. 25% Steel billet of rectangular cross-section, the width measuring less than twice the thickness, containing by weight less than 0. 25% of carbon		25	

Exhibit 1. (Continued)

序号 No.	EX①	税则号列 Tariff No.	商品名称（简称）Product Name	出口税率（%）Export Duty Rate (%)	2010年暂定税率(%) 2010 Interim Export Duty Rate (%)	2010年特别出口税率（%）2010 Special Export Duty Rate (%)
223		72071200	其他矩形截面钢坯，C，0.25% Other steel billet of rectangular cross-section, containing by weight less than 0.25% of carbon		25	
224		72071900	其他含碳量＜0.25%的钢坯 Other steel billet, containing by weight 0.25% or more of carbon		25	
226		72131000	带有轧制花纹的热轧盘条 Bars, rods, and coils, hot-rolled, containing rolling patterns		15	
227		72132000	其他易切削钢制热轧盘条 Bars, rods, and coils, hot-rolled, other, of free-cutting steel		15	
228		72139100	直径＜14mm圆截面的其他热轧盘条 Bars, rods, and coils, hot-rolled, other, of circular cross-section measuring less than 14mm in diameter		15	
229		72139900	其他热轧盘条 Bars, rods, and coils, hot-rolled, other		15	

Exhibit 1. (Continued)

序号 No.	EX ①	税则号列 Tariff No.	商品名称（简称） Product Name	出口税 率 (%) Export Duty Rate (%)	2010年暂定税率(%) 2010 Interim Export Duty Rate (%)	2010年特别 出口税率（%） 2010 Special Export Duty Rate (%)
230		72142000	热加工带有轧制花纹的条、杆 Bars and rods, thermal processed, containing rolling patterns		15	
231		72143000	热加工易切削钢的条、杆 Bars and rods, thermal processed, of free-cutting steel		15	
232		72149100	热加工其他矩形截面的条杆 Bars and rods, thermal processed, other, of rectangular cross- section		15	
233		72149900	热加工其他条、杆 Bars and rods, thermal processed, other		15	
234		72151000	冷加工其他易切削钢制条、杆 Bars and rods, cold processed, of free-cutting steel		15	
235		72155000	冷加工或冷成形的其他条、杆 Bars and rods, other, cold-formed or cold-finished		15	
236		72159000	铁及非合金钢的其他条、杆 Other bars and rods of iron or non-alloy steel		15	
237		72181000	不锈钢锭及其他初级形状产品 Stainless steel in ingots or other primary forms		15	
238		72189100	矩形截面的不锈钢半制成品 Semi-finished products of stainless steel, of rectangular cross- section		15	

Exhibit 1. (Continued)

序号 No.	EX ①	税则号列 Tariff No.	商品名称（简称）Product Name	出口税率（%）Export Duty Rate (%)	2010年暂定税率(%) 2010 Interim Export Duty Rate (%)	2010年特别出口税率（%）2010 Special Export Duty Rate (%)
239		72189900	其他不锈钢半制成品 Semi-finished products of stainless steel, other		15	
240		72191312	厚度在3毫米及以上，但小于4.75毫米的未经酸洗的按重量计含锰量在5.5%以上的铬锰系不锈钢卷板 Ferro-chromium-manganese steel in coils, of a thickness of 3mm or more but less than 4.75mm, not acid pickled, containing by weight no less than 5.5% of manganese		10	
241		72191322	厚度在3毫米及以上，但小于4.75毫米的经酸洗的按重量计含锰量在5.5%以上的铬锰系不锈钢卷板 Ferro-chromium-manganese steel in coils, of a thickness of 3mm or more but less than 4.75mm, acid pickled, containing by weight no less than 5.5% of manganese		10	

Exhibit 1. (Continued)

序号 No.	EX ①	税则号列 Tariff No.	商品名称（简称）Product Name	出口税率（%）Export Duty Rate (%)	2010年暂定税率(%) 2010 Interim Export Duty Rate (%)	2010年特别出口税率（%）2010 Special Export Duty Rate (%)
242		72191412	厚度小于3毫米的未经酸洗的按重量计含锰量在5.5%以上的铬锰系不锈钢卷板 Ferro-chromium-manganese steel in coils, of a thickness of less than 3mm, not acid pickled, containing by weight no less than 5.5% of manganese		10	
243		72191422	厚度小于3毫米的经酸洗的按重量计含锰量在5.5%以上的铬锰系不锈钢卷板 Ferro-chromium-manganese steel in coils, of a thickness of less than 3mm, acid pickled, containing by weight no less than 5.5% of manganese		10	
244		72241000	合金钢锭及其他初级形状合金钢 Other alloy steel in ingots or other primary forms		15	
245		72249010	单重≥10吨的粗铸锻件坯 Raw casting forging stocks, individual piece of a weight of 10t or more		15	
246		72249090	其他合金钢坯 Other alloy steel billets		15	

Exhibit 1. (Continued)

序号 No.	EX ①	税则号列 Tariff No.	商品名称（简称） Product Name	出口税率（%） Export Duty Rate (%)	2010年暂定税率(%) 2010 Interim Export Duty Rate (%)	2010年特别出口税率（%） 2010 Special Export Duty Rate (%)
247		74010000	铜锍、沉积铜（泥铜）Copper mattes, cement copper (precipitated copper)		15	
248		74020000	未精炼铜，电解精炼用的铜阳极 Unrefined copper, copper anodes for electrolytic refining	30	15	
249		74031111	高纯阴极铜（铜含量高于99.9935%）High-purity copper cathodes (containing by weight more than 99. 9935% of copper)	30	5	
250		74031119	其他阴极精炼铜 Other refined copper cathodes	30	10	
251		74031190	其他精炼铜的阴极型材 Other refined copper sections of cathodes	30	10	
252		74031200	精炼铜的线锭 Refined copper, wire-bars	30	10	
253		74031300	精炼铜的坯段 refined copper, billets	30	10	
254		74031900	其他未锻轧的精炼铜 Other refined copper, unwrought	30	10	
255		74032100	未锻轧的铜锌合金（黄铜）Copper-zinc base alloys (brass), unwrought	30	5	
256		74032200	未锻轧的铜锡合金（青铜）Copper-tin base alloys (bronze), unwrought	30	5	

Exhibit 1. (Continued)

序号 No.	EX ①	税则号列 Tariff No.	商品名称（简称）Product Name	出口税率(%) Export Duty Rate (%)	2010年暂定税率(%) 2010 Interim Export Duty Rate (%)	2010年特别出口税率(%) 2010 Special Export Duty Rate (%)
257		74032900	未锻轧的其他铜合金 Other copper alloys, unwrought	30	5	
258		74040000	铜废碎料 Copper waste and scrap	30	15	
259		74050000	铜母合金 Master alloys of copper		10	
260		74071000	精炼铜条、杆及型材及异型材 Copper bars, rods, and profiles, of refined copper	30	0	
261		74072100	铜锌合金条、杆及型材及异型材 Copper bars, rods, and profiles, of copper-zinc base alloys	30	0	
262		74072900	其他铜合金条杆、型材及异型材 Copper bars, rods, and profiles, of other copper alloys	30	0	
263		74081100	最大截面尺寸＞6mm的精炼铜丝 Copper wire, of refined copper, of which the maximum cross-sectional dimension exceeds 6mm	30	0	
264		74081900	其它精炼铜丝 Copper wire, of refined copper, other	30	0	

Exhibit 1. (Continued)

序号 No.	EX ①	税则号列 Tariff No.	商品名称（简称）Product Name	出口税率（%）Export Duty Rate (%)	2010年暂定税率(%) 2010 Interim Export Duty Rate (%)	2010年特别出口税率（%）2010 Special Export Duty Rate (%)
265		74082100	铜锌合金丝 Copper wire, of copper-zinc base alloys	30	0	
266		74082200	铜镍合金丝或铜镍锌合金丝 Copper wire, of copper-nickel base alloys or copper-nickel-zinc base alloys	30	0	
267		74082900	其他铜合金丝 Copper wire, of copper alloys, other	30	0	
268		74091110	含氧量不超过 10PPM的成卷的精炼铜板、片、带 Copper plates, sheets, and strips, of refined copper, in coils, containing oxygen of not more than 10PPM	30	0	
269		74091190	其他成卷的精炼铜板、片、带 Copper plates, sheets, and strips, of refined copper, in coils, other	30	0	
270		74091900	其他精炼铜板、片、带 Copper plates, sheets, and strips, of refined copper, other	30	0	
271		74092100	成卷的铜锌合金板、片、带 Copper plates, sheets, and strips, of copper-zinc base alloys, in coils	30	0	

Exhibit 1. (Continued)

序号 No.	EX ①	税则号列 Tariff No.	商品名称（简称）Product Name	出口税率(%) Export Duty Rate (%)	2010年暂定税率(%) 2010 Interim Export Duty Rate (%)	2010年特别出口税率(%) 2010 Special Export Duty Rate (%)
272		74092900	其他铜锌合金板、片、带 Copper plates, sheets, and strips, of copper-zinc base alloys, other	30	0	
273		74093100	成卷的铜锡合金板、片、带 Copper plates, sheets, and strips, of copper-tin base alloys, in coils	30	0	
274		74093900	其他铜锡合金板、片、带 Copper plates, sheets, and strips, of copper-tin base alloys, other	30	0	
275		74094000	铜镍合金或铜镍锌合金板、片、带 Copper plates, sheets, and strips, of copper-nickel base alloys or copper-nickel-zinc base alloys	30	0	
276		74099000	其他铜合金板、片、带 Copper plates, sheets, and strips, of other copper alloys	30	0	
277		75021010	高纯镍（镍含量大于99.99%，钴含量不大于0.005%）High-purity nickel (containing 99.99% or more of nickel by weight, but less than 0.005% of cobalt by weight)	40	5	
278		75021090	未锻轧的非合金镍 Unwrought nickel, not alloyed, other	40	15	
279		75022000	未锻轧镍合金 Unwrought nickel, alloys	40	15	
280		75030000	镍废碎料 Nickel waste and scrap		10	
281		75089010	电镀用镍阳极 Electroplating nickel anodes	40	15	

Exhibit 1. (Continued)

序号 No.	EX ①	税则号列 Tariff No.	商品名称（简称） Product Name	出口税率（%）Export Duty Rate (%)	2010年暂定税率(%) 2010 Interim Export Duty Rate (%)	2010年特别出口税率（%）2010 Special Export Duty Rate (%)
282		76011010	按重量计含铝量在99. 95%及以上的非合金铝 Aluminum, not alloyed, containing 99. 95% or more of aluminum by weight	30	0	
283		76011090	按重量计含铝量在99. 95%以下的非合金铝 Aluminum, not alloyed, containing less than 99. 95% of aluminum by weight	30	15	
284		76012000	未锻轧铝合金 Aluminum alloys, unwrought	30	15	
285		76020000	铝废碎料 Aluminum waste and scrap	30	15	
286		76041010	非合金铝条、杆 Aluminum bars and rods, not alloyed	20	15	
287		76041090	非合金铝型材及异型材 Aluminum profiles, not alloyed	20	0	
288		76042100	铝合金制空心异型材 Aluminum alloys hollow profiles	20	0	
289	ex	76042910	截面周长大于等于210毫米的铝合金制条、杆 Aluminum bars and rods , of aluminum alloys, with a cross-sectional perimeter of 210mm or more	20	15	

Exhibit 1. (Continued)

序号 No.	EX ①	税则号列 Tariff No.	商品名称（简称）Product Name	出口税率（%） Export Duty Rate (%)	2010年暂定税率(%) 2010 Interim Export Duty Rate (%)	2010年特别出口税率（%） 2010 Special Export Duty Rate (%)
290	ex	76042910	截面周长小于210毫米的铝合金条杆 Aluminum bars and rods, of aluminum alloys, with a cross-sectional perimeter of less than 210mm	20	5	
291		76042990	铝合金制其他型材及异型材 Aluminum profiles, of aluminum alloys, other	20	0	
292		76051100	最大截面尺寸超过7mm的非合金铝丝 Aluminum wire, of aluminum, not alloyed	20	0	
293		76051900	其他非合金铝丝 Aluminum wire, of aluminum, not alloyed, other	20	0	
294		76052100	最大截面尺寸超过7mm的铝合金丝 Aluminum wire, of aluminum alloys, of which the maximum cross-sectional dimension exceeding 7mm	20	0	
295		76052900	其他铝合金丝 Aluminum wire, other	20	0	

Exhibit 1. (Continued)

序号 No.	EX ①	税则号列 Tariff No.	商品名称（简称） Product Name	出口税率(%) Export Duty Rate (%)	2010年暂定税率(%) 2010 Interim Export Duty Rate (%)	2010年特别出口税率（%） 2010 Special Export Duty Rate (%)
296		76061120	厚度在0.3mm及以上，但不超过0.36mm的非合金铝制矩形铝板片带 Aluminum plates, sheets, and strips, rectangular, of aluminum, not alloyed, of a thickness of 0.30mm or more but not exceeding 0.36mm	20	0	
297		76061190	非合金铝制矩形的其他板、片及带 Aluminum plates, sheets, and strips, rectangular, of aluminum, not alloyed, other	20	0	
298		76061220	厚度 <0.28mm的铝合金制矩形铝板片带 Aluminum plates, sheets, and strips, rectangular, of aluminum alloys, of athickness of less than 0.28mm	20	0	
299		76061230	厚度在0.28mm及以上，但不超过0.35mm的铝合金制矩形铝板片带 Aluminum plates, sheets, and strips, rectangular, of aluminum alloys, of a thickness of 0.28mm or more but not exceeding 0.35mm	20	0	
300		76061250	0.35mm<厚度≤0.4mm的铝合金制矩形铝板片带 Aluminum plates, sheets, and strips, rectangular, of aluminum alloys, of a thickness of more than 0.35mm but not exceeding 0.4mm	20.	0	

Exhibit 1. (Continued)

序号 No.	EX ①	税则号列 Tariff No.	商品名称（简称）Product Name	出口税率（%）Export Duty Rate (%)	2010年暂定税率(%) 2010 Interim Export Duty Rate (%)	2010年特别出口税率（%）2010 Special Export Duty Rate (%)
301		76061290	厚度>0.4mm的铝合金制矩形铝板片带 Aluminum plates, sheets, and strips, rectangular, of aluminum alloys, of a thickness of more than 0.4mm	20	0	
302		76069100	非合金铝制非矩形的板、片及带 Aluminum plates, sheets, and strips, not rectangular, of aluminum, not alloyed	20	0	
303		76069200	铝合金制非矩形的板、片及带 Aluminum plates, sheets, and strips, not rectangular, of aluminum alloys	20	0	
304		78011000	未锻轧精炼铅 Unwrought refined lead		10	
305		78020000	铅废碎料 Lead waste and scrap		10	
306		79011110	按重量计含锌量在99.995%及以上的未锻轧锌 Unwrought zinc, containing by weight 99.995% or more of zinc	20	0	

Exhibit 1. (Continued)

序号 No.	EX ①	税则号列 Tariff No.	商品名称（简称）Product Name	出口税率 (%) Export Duty Rate (%)	2010年暂定税率(%) 2010 Interim Export Duty Rate (%)	2010年特别出口税率（%）2010 Special Export Duty Rate (%)
307		79011190	99.99≤含锌量<99.995%的未锻轧锌 Unwrought zinc, containing by weight 99.99% or more, but less than 99.995% of zinc	20	5	
308		79011200	含锌量<99.99%的未锻轧锌 Unwrought zinc, containing by weight less than 99.99% of zinc	20	15	
309		79012000	未锻轧锌合金 Unwrought zinc alloys	20	0	
310		79020000	锌废碎料 Zinc waste and scrap		10	
311		80011000	非合金锡 Tin, not alloyed		10	
312		80020000	锡废碎料 Tin waste and scrap		10	
313		81011000	钨粉 Tungsten powders		5	
314		81019400	未锻轧钨 Unwrought tungsten		5	
315		81019700	钨废碎料 Tungsten waste and scrap		15	
316		81021000	钼粉 Molybdenum powders		5	
317		81029400	未锻轧钼 Unwrought molybdenum		5	
318		81029700	钼废碎料 Molybdenum waste and scrap		15	

Exhibit 1. (Continued)

序号 No.	EX ①	税则号列 Tariff No.	商品名称（简称）Product Name	出口税率 (%) Export Duty Rate (%)	2010年暂定税率(%) 2010 Interim Export Duty Rate (%)	2010年特别出口税率（%）2010 Special Export Duty Rate (%)
319		81033000	钽废碎料 Tantalum waste and scrap		10	
320		81041100	按重量计含镁量≥99.8%的未锻轧镁 Unwrought magnesium, containing 99.8% or more of magnesium by weight		10	
321		81041900	其他未锻轧镁 Other unwrought magnesium		10	
322		81042000	镁废碎料 Magnesium waste and scrap		10	
323		81101010	未锻轧锑 Unwrought antimony	20	5	
324		81101020	锑粉末 Antimony powders	20		
325		81102000	锑废碎料 Antimony waste and scrap	20		
326		81110010	未锻轧锰、锰废碎料;粉末 Unwrought manganese, manganese waste and scrap, powders		20	
327		81122100	未锻轧铬、铬粉末 Unwrought chromium, chromium powders		15	
328		81122200	铬废碎料 Chromium waste and scrap		15	
329		81129230	未锻轧铟;铟废碎料;铟粉末 Unwrought indium, indium waste and scrap, indium powders		5	

注① : "ex"表示应税商品范围以"商品名称"描述为准, 其余以税号为准。

注② : 出口价格包括海关认可的货物货价, 货物运至中华人民共和国境内输出地点装载前的运输及其相关费用, 保险费。 淡季

当出口价格高于基准价格时, 税率计算结果四舍五入保留3位小数。

Note:

① "Ex" means the scope of the product covered by the tariff number is governed by the description in the column of "product name."

Others are governed by the tariff number.

② Export price includes product price, transportation and related expenses and insurance incurred prior to arrival and loading at the export location within the PRC, recognized by the Customs. If export price exceeds base price during off season, the duty rate calculation result should be rounded to three decimal places.

Exhibit 2. 2010 1st Batch Export Quota Distribution for Rare-Earth Materials in General Trade*

Translator's note: China seems to distinguish between general trade, processing trade, border trade, and aid trade.

序号 No.	公司名称 Company Name	配额数量（吨） Quota (MT)
	合计 Total	**16304**
1	中国中钢集团公司 China Sinosteel Corporation	784
2	中国五矿集团公司 China Minmetals Corporation	1182
3	中国有色金属进出口江苏公司 China National Nonferrous Metals Import & Export Jiangsu Corp.	726
4	广东广晟有色金属进出口有限公司 China National Nonferrous Metals Import & Export Guangdong Co., Ltd.	518
5	常熟盛昌稀土冶炼厂 Changshu Shengchang Rare Earth Smeltery	337
6	江苏卓群纳米稀土股份有限公司 Jiangsu GeoQuin Nano Rare Earth Co., Ltd.	248
7	江西金世纪新材料股份有限公司 Jiangxi Golden Century Advanced Materials Co., Ltd.	374
8	内蒙古包钢和发稀土有限公司 Inner Mongolia Hefa Rare Earth Science & Technology Development Co., Ltd.	1504
9	江西南方稀土高技术股份有限公司 Jiangxi South Rare Earth Hi-Tech Co., Ltd.	629
10	赣州晨光稀土新材料有限公司 Ganzhou Chenguang Rare Earth Advanced Material Co., Ltd.	601
11	赣州虔东稀土集团股份有限公司 Ganzhou Qiandong Rare Earth Group Co., Ltd.	536
12	有研稀土新材料股份有限公司 Grirem Advanced Materials Co., Ltd.	469
13	益阳鸿源稀土有限责任公司 Yiyang Hongyuan Rare Earth Co., Ltd.	837

Exhibit 2. (Continued)

序号 No.	公司名称 Company Name	配额数量（吨） Quota (MT)
14	包头华美稀土高科有限公司 Baotou Huamei Rare-Earth Hi-Tech Co., Ltd.	1659
15	内蒙古包钢稀土(集团)高科技股份有限公司Inner Mongolia Baotou Steel Rare-Earth (Group) Hi-Tech Co., Ltd.	1350
16	甘肃稀土新材料股份有限公司 Gansu Rare Earch New Materials LLC	1069
17	乐山盛和稀土科技有限公司 Leshan Shenghe Rare Earth Technology Co., Ltd.	1102
18	阜宁稀土实业有限公司 Funing Rare Earth Industrial Co., Ltd.	477
19	山东鹏宇实业股份有限公司 Shandong Pengyu Industrial Co., Ltd.	986
20	赣县红金稀土有限公司 Ganxian Hongjin Rare Earth Co., Ltd.	239
21	徐州金石彭源稀土材料厂 Xuzhou Jinshi Pengyuan Rare Earth Materials Factory	374
22	广东珠江稀土有限公司 Guangdong Zhujiang Rare Earth Co., Ltd.	304

某企业应得配额 = 此次下达配额量× 〔0.78×（A1+A2）＋0.22×A3〕

A1 =（各企业近三年出口数量÷全国出口总量）×0.9权重

A2 =（各企业近三年出口金额÷全国出口金额）×0.1权重

A3 = 生产企业2008年出口供货量÷各生产企业出口供货总量

A company's quota = total quota x [0.78 x (A1 + A2) + 0.22 x A3]

A1 = (the company's export quantity in the most recent 3 years ÷ entire country's export quantity) x 0.9 weighted average

A2 = (the company's export revenue in the most recent 3 years ÷ entire country's export revenue) x 0.1 weighted average

A3 = the manufacturing company's 2008 export quantity ÷ all manufacturing companies' export quantity

Exhibit 3. Notice on Application Criteria and Procedures for 2010 Rare-Earth Materials Export Quota

Summary of Key Points
1. This does not apply to foreign-invested enterprises.
2. A manufacturing enterprise must: (1) Register with Industrial and Commercial Administrative Department, obtain import and export operational qualification or register as a foreign trading enterprise, and have independent legal personality. Enterprises registered after 2005 must be approved by national supervising department. (2) Comply with rare earth industry planning, policy, and management, and have exported either 2,000 MT or more products or have export revenue of RMB 70,000,000 or more in 2008. (3) If the conditions set in the subparagraph immediately above are not met, the manufacturing enterprise must have, in the most recent 3 years (2006-2008), an average annual export quantity of 1,500 MT or more, or average annual export revenue of USD 15,000,000 or more (the numbers will be based on Custom's statistics). (4) Product quality must meet current national standards, and obtain ISO9000 quality system certification.
(5) The manufacturing enterprise's rare earth materials must come from rare earth mining enterprises that have mining qualifications as proclaimed by the Ministry of Land and Resources.
(6) Have environmental protection facilities (including online monitoring facilities) that are compatible with the scale of production, pollutants emission is in compliance with national or local standards, have timely paid emission fees in full in 2008 and 2009 as verified by the environmental protection department at the provincial level or above, have not violated environmental laws, have formulated environmental emergency plan, and have complete and sound supporting facilities. (7) Comply with national land management policies and regulations. (8) Comply with national laws and regulations and relevant local regulations, lawfully participate in pension, unemployment, medical care, work injury, maternity, and other social insurance, timely pay social insurance premiums in full, and provide proof of the same from local labor and social insurance departments. (9) Have not violated national laws and regulations.
3. A trading enterprise [note: literal translation would be "circulation enterprise"] must: (1) Register with Industrial and Commercial Administrative Department, obtain import and export operational qualification or register as a foreign trading enterprise, and have independent legal personality. (2) Have registered capital of RMB 50,000,000 or more, and have, in the most recent 3 years (2006-2008), an average annual export quantity of 1,500 MT or more, or average annual export revenue of USD 15,000,000 or more (the numbers will be based on Custom's statistics). (3) Its export products must come from manufacturing enterprises in compliance with Paragraph 2, Subparagraphs (1), (4)-(9), and it must provide relevant documents on its supply manufacturing enterprises' such compliance, the related value added tax invoices, and proof of its supply enterprises. (4) Comply with national laws and regulations and relevant local regulations, lawfully participate in pension, unemployment, medical care, work injury, maternity, and other social insurance, timely pay social insurance premiums in full, and provide proof of the same from local labor and social insurance departments. (5) Obtain ISO9000 quality system certification. (6). Have not violated national laws and regulations.

Exhibit 3. (Continued)

4. In order to concentrate production and decrease the number of export enterprises, the export performance of parents, subsidiaries, and affiliates of companies that are registered after January 1, 2007 will not be recognized.
5. If an enterprise violates laws and regulations or environmental protection policies after receiving the quota, its quota will be withdrawn, suspended, or cancelled once its violation is confirmed.
6. Application and Approval Procedures: (1) Local rare earth export enterprises shall submit their applications to provincial level commercial supervising department. The provincial level commercial supervising department will conduct an initial verification under the application criteria set forth above, submit its decision and selected list of qualified enterprises to the Ministry of Commerce (MOFCOM), and copy the same to China Chamber of Commerce of Metals Minerals & Chemicals Importers & Exporters (CCMMC) by November 20, 2009. (2) Rare earth export enterprises managed by central government shall submit their application directly to the MOFCOM, and copy the same to CCMMC.
(3) MOFCOM commissions CCMMC to conduct a review of the provincial level verification. CCMMC shall submit its decision to MOFCOM by November 27, 2009. (4) MOFCOM will decide and publish the final list of companies based on CCMMC's review decision.
7. Basically, both manufacturing and trading enterprises have to submit supporting documents on every item listed in the criteria.

Exhibit 4. Excerpt from the U.S.-China Economic and Security Review Commission (USCC) 2009 Report to Congress, pp. 62-63 available at http://www.uscc.gov/annualreport/2009/annualreportfull09.pdf

Export Restrictions

Export restrictions or export quotas, especially on energy and raw materials, have two general effects: First, they suppress prices in the domestic market for these goods, which lowers production costs for industries that use the export-restricted materials; and second, these restrictions increase the world price for the raw materials that are affected by limiting the world supply, thereby raising production costs in competing countries.[1]

According to the USTR, "despite China's commitment since its accession to the WTO to eliminate all taxes and charges on exports, including export duties . . . China has continued to impose restrictions on exports of certain raw materials,[2] including quotas, related licensing requirements, and duties, as China's state planners have continued to guide the development of downstream industries."[3] The USTR's 2009 report on foreign trade barriers concludes that

"China's export restrictions affect U.S. and other foreign producers on a wide range of downstream products such as steel, chemicals, ceramics, semiconductor chips, refrigerants, medical imagery, aircraft, refined petroleum products, fiber optic cables, and catalytic converters, among many others."[4]

In June 2009, the Obama Administration initiated a WTO case against China over export restraints on numerous important raw materials. U.S. officials have been concerned for years about export restraints on raw materials from China and, in cooperation with European and Japanese officials, have held regular bilateral and multilateral discussions with Chinese officials since China joined the WTO, before the WTO's Import Licensing Committee.[5] The USTR reports that these efforts had no effect and that China in fact increased export restraints on raw materials over time.[6] According to the USTR, "China's measures appear to be part of a troubling industrial policy aimed at providing a substantial competitive advantage for the Chinese industries using these inputs."[7] Others have reported concerns that China's export restrictions are part of a larger effort to stockpile resources in order to insulate China from sudden fluctuations in global commodities markets and to increase China's ability to influence those markets.[8]

China's Restrictions on Exports of Rare Earth Minerals

China appears to be tightening its control over the supply of rare earth elements, valuable minerals that are used prominently in the production of such high-technology goods as flat panel screens and cell phones, and crucial green technologies such as hybrid car batteries and the special magnets used in wind turbines.[9] Rare earth minerals are also critical for many military technologies, including the magnets used in the guidance systems of U.S. military smart bombs like Joint Direct Attack Munitions, and super-alloys (used to make parts for jet aircraft engines).

China accounts for the vast majority—93 percent—of the world's production of rare earth minerals, and for the last three years it has been reducing the amount that can be exported.[10] After a draft policy outlining the tightening of exports for rare earth minerals was issued in August 2009 by the Ministry of Industry and Information Technology, Zhao Shuanglian, deputy chief of the Inner Mongolia autonomous region, spoke out to quell global concerns. According to Mr. Zhao, rare earth elements are "the most important resource for Inner Mongolia," which contains 75 percent of China's deposits, and by cutting exports and Controlling production, the government wants to "attract users of rare earths to set up in Inner

Mongolia" to develop manufacturing.[11] China also is taking steps to consolidate its rare earths industry, with the aim of creating a consortium of miners and processors in Inner Mongolia.[12]

China's Ministry of Industry and Information Technology says it is limiting production in some mines and closing others completely because some of the rare earths are extracted under dire environmental conditions, but tighter limits on exports of rare earths place foreign manufacturers at a disadvantage compared to the domestic producers, whose access will not be so restricted. There has been no official U.S. government response so far, but a spokeswoman for the U.S. embassy in Beijing questioned the WTO-legality of such restrictions, noting that "[w]e would be concerned by any WTO member's policies that appear to be inconsistent with its WTO obligations."[13]

End Notes

[1] Peter Navarro, "Benchmarking the Advantages Foreign Nations Provide their Manufacturers," in Richard McCormack, ed., *Manufacturing a Better Future for America* (Washington, DC: The Alliance for American Manufacturing, 2009), 113.

[2] China maintains export quotas and, at times, export duties on antimony, bauxite, coke, fluorspar, indium, magnesium carbonate, molybdenum, rare earths, silicon, talc, tin, tungsten, and zinc.

[3] U.S. Trade Representative, *2009 National Trade Estimate Report on Foreign Trade Barriers* (Washington, DC: March 29, 2009), 97.

[4] U.S. Trade Representative, *2009 National Trade Estimate Report on Foreign Trade Barriers* (Washington, DC: March 29, 2009), 97.

[5] U.S. Trade Representative, *2008 Report to Congress on China's WTO Compliance* (Washington, DC: December 2008), 29-30.

[6] U.S. Trade Representative, *2008 Report to Congress on China's WTO Compliance* (Washington, DC: December 2008), 5-6.

[7] *USTR News*, "United States Files WTO Case Against China Over Export Restraints on Raw Materials" (Washington, DC: U.S. Trade Representative, June 23, 2009).

[8] STRATFOR, "China: Alleged WTO Violations and Commodity Prices" (Austin, TX: June 24, 2009).

[9] The rare earth elements group is comprised of 17 minerals—scandium, yttrium, and the 15 lanthanoids—and they play a crucial role in many advanced technological devices. For more information, see, for example, Gordon B. Haxel, James B. Hedrick, and Greta J. Orris, "Rare Earth Elements—Critical Resources for High Technology," U.S. Geological Survey Fact Sheet 087-02 (Reston, VA: 2002). http://pubs.usgs.gov/fs/2002/fs087–02/fs087–02.pdf.

[10] Keith Bradsher, "China Tightens Grip on Rare Minerals," *New York Times*, September 1, 2009.

[11] Xiao Yu and Eugene Tang, "China Considers Rare-Earth Reserve in Inner Mongolia," Bloomberg, September 2, 2009.

[12] Chuin-Wei Yap, "China Plays Down Rare Earth Fears," *Wall Street Journal*, September 2, 2009.

[13] Keith Bradsher, "Backpedaling, China Eases Proposal to Ban Exports of Some Vital Minerals," *New York Times*, September 3, 2009. http://www.nytimes.com/2009/09/04/business/global/04minerals.html.

In: Rare Earth Minerals: Policies and Issues   ISBN: 978-1-61122-310-1
Editor: Steven M. Franks   © 2011 Nova Science Publishers, Inc.

Chapter 10

RARE EARTHS[1]

U.S. Geological Survey, Mineral Commodity Summaries

Domestic Production and Use: In 2009, rare earths were not mined in the United States; however, rare-earth concentrates previously produced at Mountain Pass, CA, were processed into lanthanum concentrate and didymium (75% neodymium, 25% praseodymium) products. Rare-earth concentrates, intermediate compounds, and individual oxides were available from stocks. The United States continued to be a major consumer, exporter, and importer of rare-earth products in 2009. The estimated value of refined rare earths imported by the United States in 2009 was $84 million, a decrease from $186 million imported in 2008. Based on final 2008 reported data, the estimated 2008 distribution of rare earths by end use, in decreasing order, was as follows: metallurgical applications and alloys, 29%; electronics, 18%; chemical catalysts, 14%; rare-earth phosphors for computer monitors, lighting, radar, televisions, and x-ray-intensifying film, 12%; automotive catalytic converters, 9%; glass polishing and ceramics, 6%; permanent magnets, 5%; petroleum refining catalysts, 4%; and other, 3%.

Salient Statistics—United States:	2005	2006	2007	2008	2009[e]
Production, bastnäsite concentrates	—	—	—	—	—
Imports:[2]					
Thorium ore (monazite or various thorium materials)	—	—	—	—	20
Rare-earth metals, alloy	880	867	784	679	210
Cerium compounds	2,170	2,590	2,680	2,080	1,190
Mixed REOs	640	1,570	2,570	2,390	2,760
Rare-earth chlorides	2,670	2,750	1,610	1,310	390
Rare-earth oxides, compounds	8,550	10,600	9,900	8,740	2,160
Ferrocerium, alloys	130	127	123	125	100
Exports:[2]					
Thorium ore (monazite or various thorium materials)	—	—	1	61	23
Rare-earth metals, alloys	636	733	1,470	1,390	6,500
Cerium compounds	2,210	2,010	1,470	1,380	690
Other rare-earth compounds	2,070	2,700	1,300	663	420
Ferrocerium, alloys	4,320	3,710	3,210	4,490	2,540
Consumption, apparent (excludes thorium ore)	6,060	9,350	10,200	7,410	([3])
Price, dollars per kilogram, yearend:					
Bastnäsite concentrate, REO basis[e]	5.51	6.06	6.61	8.82	5.73
Monazite concentrate, REO basis[e]	0.54	0.87	0.87	0.87	0.87
Mischmetal, metal basis, metric ton quantity[4]	5-6	5-6	7-8	8-9	8-9
Stocks, producer and processor, yearend	W	W	W	W	W
Employment, mine and mill, number at yearend	71	65	70	100	110
Net import reliance[5] as a percentage of apparent consumption	100	100	100	100	100

Recycling: Small quantities, mostly permanent magnet scrap.

Import Sources (2005-08): Rare-earth metals, compounds, etc.: China, 91%; France, 3%; Japan, 3%; Russia, 1%; and other, 2%.

Tariff: Item	Number	Normal Trade Relations 12-31 -09
Thorium ores and concentrates (monazite)	2612.20.0000	Free.
Rare-earth metals, whether or not intermixed or interalloyed	2805.30.0000	5.0% ad val.
Cerium compounds	2846.10.0000	5.5% ad val.
Mixtures of REOs (except cerium oxide)	2846.90.2010	Free.
Mixtures of rare-earth chlorides (except cerium chloride)	2846.90.2050	Free.
Rare-earth compounds, individual		
REOs (excludes cerium compounds)	2846.90.8000	3.7% ad val.
Ferrocerium and other pyrophoric alloys	3606.90.3000	5.9% ad val.

Depletion Allowance: Monazite, 22% on thorium content and 14% on rare-earth content (Domestic), 14% (Foreign); bastnäsite and xenotime, 14% (Domestic and foreign).

Government Stockpile: None.
Prepared by James B. Hedrick [(703) 648-7725, jhedrick@usgs.gov, fax: (703) 648-7757]

Events, Trends, and Issues: Domestic consumption of rare earths in 2009 decreased substantially, based on apparent consumption (derived from 8 months of trade data). Only one of seven rare-earth import categories increased when compared with those of 2008—the category "mixtures of REOs (except cerium oxide)." Prices were generally lower in 2009 compared with those of 2008 for most rare-earth products amid decreased consumption and a declining supply. Consumption for most rare-earth uses in the United States decreased as a consequence of the worldwide economic downturn. The economic downturn lowered consumption of cerium compounds used in automotive catalytic converters and in glass additives and glass-polishing compounds; rare-earth chlorides used in the production of fluid-cracking catalysts used in oil refining; rare-earth compounds used in automotive catalytic converters and many other applications; rare-earth metals and their alloys used in armaments, base-metal alloys, lighter flints, permanent magnets, pyrophoric alloys, and superalloys;

yttrium compounds used in color televisions and flat-panel displays, electronic thermometers, fiber optics, lasers, and oxygen sensors; and phosphors for color televisions, electronic thermometers, fluorescent lighting, pigments, superconductors, x-ray-intensifying screens, and other applications. The trend is for a continued increase in the use of rare earths in many applications, especially automotive catalytic converters, permanent magnets, and rechargeable batteries for electric and hybrid vehicles.

The rare-earth separation plant at Mountain Pass, CA, resumed operations in 2007 and continued to operate throughout 2009. Bastnäsite concentrates and other rare-earth intermediates and refined products continued to be sold from mine stocks at Mountain Pass. Exploration for rare earths continued in 2009; however, global economic conditions were not as favorable as in early 2008. Economic assessments continued at Bear Lodge in Wyoming; Diamond Creek in Idaho; Elk Creek in Nebraska; Hoidas Lake in Saskatchewan, Canada; Nechalacho (Thor Lake) in Northwest Territories, Canada; Kangankunde in Malawi; Lemhi Pass in Idaho-Montana; Nolans Project in Northern Territory, Australia; and various other locations around the world. At the Mount Weld rare-earth deposit in Australia, the initial phase of mining of the open pit was completed in June 2008. A total of 773,000 tons of ore was mined at an average grade of 15.4% REO; however, no beneficiation plant existed to process the ore into a rare-earth concentrate. Based on the fine-grained character of the Mt. Weld ore, only 50% recovery of the REO was expected.

World Mine Production and Reserves: Reserves data for Australia, China, and India were updated based on data from the respective countries.

	Mine production[e] 2008	2009	Reserves[6]
United States	——	——	13,000,000
Australia	——	——	5,400,000
Brazil	650	650	48,000
China	120,000	120,000	36,000,000
Commonwealth of Independent States	NA	NA	19,000,000
India	2,700	2,700	3,100,000
Malaysia	380	380	30,000
Other countries	NA	NA	22,000,000
World total (rounded)	**124,000**	**124,000**	**99,000,000**

World Resources: Rare earths are relatively abundant in the Earth's crust, but discovered minable concentrations are less common than for most other ores. U.S. and world resources are contained primarily in bastnäsite and monazite. Bastnäsite deposits in China and the United States constitute the largest percentage of the world's rare- earth economic resources, while monazite deposits in Australia, Brazil, China, India, Malaysia, South Africa, Sri Lanka, Thailand, and the United States constitute the second largest segment. Apatite, cheralite, eudialyte, loparite, phosphorites, rare-earth-bearing (ion adsorption) clays, secondary monazite, spent uranium solutions, and xenotime make up most of the remaining resources. Undiscovered resources are thought to be very large relative to expected demand.

Substitutes: Substitutes are available for many applications but generally are less effective.

End Notes

^eEstimated. NA Not available. W Withheld to avoid disclosing company proprietary data. — Zero.

[1] Data include lanthanides and yttrium but exclude most scandium. See also Scandium and Yttrium.

[2] REO equivalent or contents of various materials were estimated. Data from U.S. Census Bureau.

[3] Without drawdown in producer stocks (withheld), apparent consumption calculations in 2009 resulted in a negative number.

[4] Price range from Elements—Rare Earths, Specialty Metals and Applied Technology, Trade Tech, Denver, CO, and Web-based High Tech Materials, Longmont, CO, and Hefa Rare Earth Canada Co. Ltd., Richmond, British Columbia, Canada.

[5] Defined as imports − exports + adjustments for Government and industry stock changes. For 2007 through 2009, excludes producer stock changes.

[6] See Appendix C for definitions. Reserve base estimates were discontinued in 2009; see Introduction.

In: Rare Earth Minerals: Policies and Issues   ISBN: 978-1-61122-310-1
Editor: Steven M. Franks   © 2011 Nova Science Publishers, Inc.

Chapter 11

YTTRIUM

U.S. Geological Survey

(Data in metric tons of yttrium oxide (Y_2O_3) content unless otherwise noted)

Domestic Production and Use: The rare-earth element yttrium was not mined in the United States in 2009. All yttrium metal and compounds used in the United States were imported. Principal uses were in phosphors for color televisions and computer monitors, temperature sensors, trichromatic fluorescent lights, and x-ray-intensifying screens. Yttria-stabilized zirconia was used in alumina-zirconia abrasives, bearings and seals, high-temperature refractories for continuous-casting nozzles, jet-engine coatings, oxygen sensors in automobile engines, simulant gemstones, and wear-resistant and corrosion-resistant cutting tools. In electronics, yttrium-iron garnets were components in microwave radar to control high-frequency signals. Yttrium was an important component in yttriumaluminum-garnet laser crystals used in dental and medical surgical procedures, digital communications, distance and temperature sensing, industrial cutting and welding, nonlinear optics, photochemistry, and photoluminescence. Yttrium also was used in heating-element alloys, high-temperature superconductors, and superalloys. The approximate distribution in 2008 by end use was as follows: phosphors (all types), 87%; ceramics, 10%; metallurgy, 2%; and electronics and lasers, 1 %.

Recycling: Small quantities, primarily from laser crystals and synthetic garnets.

Import Sources (2005-08):[e, 7] Yttrium compounds, greater than 19% to less than 85% weight percent yttrium oxide equivalent: China, 95%; Japan, 4%; France, less than one-half of 1%; and other, insignificant. Import sources based on Journal of Commerce data (2008 only): China, 90%; Austria, 8%; Japan, 1%; and United Kingdom, 1%.

Depletion Allowance: Monazite, thorium content, 22% (Domestic), 14% (Foreign); yttrium, rare-earth content, 14% (Domestic and foreign); and xenotime, 14% (Domestic and foreign).

Salient Statistics—United States:	2005	2006	2007	2008	2009[e]
Production, mine	—	—	—	—	—
Imports for consumption:					
In monazite	—	—	—	—	—
Yttrium, alloys, compounds, and metal[e,2]	582	742	676	616	400
Exports, in ore and concentrate	NA	NA	NA	NA	NA
Consumption, estimated[3]	582	742	676	616	400
Price, dollars:					
Monazite concentrate, per metric ton[4]	300	300	300	300	300
Yttrium oxide, per kilogram, 99.9% to 99.99% purity[5]	10-85	10-85	10-85	10-85	10-85
Yttrium metal, per kilogram, 99.9% purity[5]	96	68-155	68-155	68-155	68-155
Stocks, processor, yearend	NA	NA	NA	NA	NA
Net import reliance[e, 6] as a percentage of apparent consumption	100	100	100	100	100

Tariff: Item	Number	Normal Trade Relations 12-31 -09
Thorium ores and concentrates (monazite)	2612.20.0000	Free.
Rare-earth metals, scandium and yttrium, whether or not intermixed or interalloyed	2805.30.0000	5.0% ad val.
Yttrium-bearing materials and compounds containing by weight >19% to <85% Y2O3	2846.90.4000	Free.
Other rare-earth compounds, including yttrium oxide >85% Y2O3, yttrium nitrate, and other individual compounds	2846.90.8000	3.7% ad val.

Government Stockpile: None.

Events, Trends, and Issues: Estimated yttrium consumption in the United States decreased in 2008 and was expected to decrease again in 2009. The

United States required yttrium for use in phosphors and in electronics, especially those used in defense applications.

Yttrium production and marketing within China continued to be competitive, and the price range for yttrium has remained steady, although one domestic supplier lowered its oxide price from $50 to $44 per kilogram in 2009. China was the source of most of the world's supply of yttrium, from its weathered clay ion-adsorption ore deposits in the southern Provinces, primarily Fujian, Guangdong, and Jiangxi, with a lesser number of deposits in Guangxi and Hunan. Processing was primarily at facilities in Guangdong, Jiangsu, and Jiangxi Provinces. Yttrium was consumed mainly in the form of high-purity oxide compounds for phosphors. Smaller amounts were used in ceramics, electronic devices, lasers, and metallurgical applications.

China was the primary source of most of the yttrium consumed in the United States. About 90% of the imported yttrium compounds, metal, and alloys were sourced from China, with lesser amounts from Austria, Japan, and the United Kingdom.

World Mine Production and Reserves

	Mine	productione,[8]	Reserves[9]
	2008	2009	
United States	—	—	120,000
Australia	—	—	100,000
Brazil	15	15	2,200
China	8,800	8,800	220,000
India	55	55	72,000
Malaysia	4	4	13,000
Sri Lanka	—	—	240
Other countries	—	—	17,000
World total (rounded)	8,900	8,900	540,000

World Resources: Although reserves may be sufficient to satisfy near term demand at current rates of production, economics, environmental issues, and permitting and trade restrictions could affect the mining or availability of many of the rare-earth elements, including yttrium. Large resources of yttrium in monazite and xenotime are available worldwide in ancient and recent placer deposits, carbonatites, uranium ores, and weathered clay deposits (ion-adsorption ore). Additional large subeconomic resources of yttrium occur in apatite-magnetite-bearing rocks, deposits of niobium-tantalum minerals, non-placer monazite-bearing deposits, sedimentary phosphate deposits, and uranium ores, especially those of the Blind River District near Elliot Lake, Ontario, Canada, which contain yttrium in brannerite, monazite, and uraninite. Additional resources in Canada are contained in allanite, apatite, and britholite at Eden Lake, Manitoba; allanite and apatite at Hoidas Lake, Saskatchewan; and fergusonite and xenotime at Thor Lake, Northwest Territories. The world's resources of yttrium are probably very large. Yttrium is associated with most rare-earth deposits. It occurs in various minerals in differing concentrations and occurs in a wide variety of geologic environments, including alkaline granites and intrusives, carbonatites, hydrothermal deposits, laterites, placers, and vein-type deposits.

Substitutes: Substitutes for yttrium are available for some applications but generally are much less effective. In most uses, especially in electronics, lasers, and phosphors, yttrium is not subject to substitution by other elements. As a stabilizer in zirconia ceramics, yttria (yttrium oxide) may be substituted with calcia (calcium oxide) or magnesia (magnesium oxide), but they generally impart lower toughness.

End Notes

[e] Estimated. NA Not available. — Zero.

[1] See also Rare Earths.

[2] Imports based on data from the Port Import/Export Reporting Service (PIERS), Journal of Commerce.

[3] Essentially, all yttrium consumed domestically was imported or refined from imported ores and concentrates.

[4] Monazite price based on monazite exports from Malaysia for 2005 and estimated for 2006 through 2009.

[5] Yttrium oxide and metal prices for 5-kilogram to 1-metric-ton quantities from Rhodia Rare Earths, Inc., Shelton, CT; the China Rare Earth Information Center, Baotou, China; Hefa Rare Earth Canada Co., Ltd., Vancouver, Canada; and Stanford Materials Corp., Aliso Viejo, CA.
[6] Defined as imports − exports + adjustments for Government and industry stock changes.

[7] May not add to 100% due to rounding.

[8] Includes yttrium contained in rare-earth ores.

[9] See Appendix C for definitions. Reserve base estimates were discontinued in 2009; see Introduction.

CHAPTER SOURCES

The following chapters have been previously published:

Chapter 1 – This is an edited reformatted and augmented version of a Congressional Research Service publication, report R41347, dated July 28, 2010.

Chapter 2 – This is an edited reformatted and augmented version of a United States Government Accountability Office publication, report GAO-10-617R, dated April 14, 2010.

Chapter 3 - These remarks were delivered as testimony given on March 16, 2010. Chairman Bart Gordon, before the Committee on Science and Technology, United States House of Representatives.

Chapter 4 - These remarks were delivered as testimony given on March 16, 2010. Chairman Brad Miller, before the Committee on Science and Technology, United States House of Representatives.

Chapter 5 - These remarks were delivered as testimony given on March 16, 2010. Stephen Freiman, President, Freiman Consulting and Representing the Ad Hoc Committee on Critical Mineral Impacts on the United States Economy, Committee on Earth Resources, Board on Earth Sciences and Resources, National Research Council, The National Academies, before the Subcommittee on Investigations and Oversight Committee on Science and Technology, United States House of Representatives.

Chapter 6 – These remarks were delivered as responses to Subcommittee's Questions. Karl A. Gschneidner, Jr., Ames Laboratory, United States Department of Energy and Department of Materials Science and Engineering.

Chapter 7 – These remarks were delivered as testimony given on February 10, 2010. Steven J. Duclos, Chief Scientist and Manager, Material Sustainability, GE Global Research, before the Subcommittee on Investigations and Oversight of the House Committee on Science and Technology.

Chapter 8 – These remarks were delivered as testimony given on March 16, 2010. Mark A. Smith, Chief Executive Officer, Molycorp Minerals, LLC, before House Science and Technology Committee, Subcommittee on investigations and Oversight.

Chapter 9 – These remarks were delivered as testimony given on March 16, 2010. Terence P. Stewart, Managing Partner, Law Offices of Stewart and Stewart, before the United States House of Representatives Committee on Science and Technology, Subcommittee on Investigations and Oversight.

Chapter 10 – This is an edited reformatted and augmented version of a United States Geological Survey publication, survey Mineral Commodity Summaries, dated January 2010.

Chapter 11 – This is an edited reformatted and augmented version of a United States Geological Survey publication, survey Mineral Commodity Summaries, dated January 2010.

INDEX

E

F

T